Senility
Radiation Poisoning
Cancer
Muscular Dystrophy
Heart Disease
Arthritis

Death and taxes are inevitable. So, for most of us, are the conditions listed above—if they're going to happen, they'll happen, and what is there to do about it? Astonishingly, there is evidence that a rare trace element provides protection against them—and has in fact been shown to reverse some of them. That element is *selenium.* It occurs in miniscule traces in plants and in the animals which feed on those plants—except where it no longer exists in the soil. Dr. Passwater explains how we can add to our bodies' store of this vital element, and in what ways it can strengthen our natural defenses against degenerative and acute diseases.

Selenium as Food & Medicine

Richard A. Passwater, Ph.D.

Introduction by Julian E. Spallholz, Ph.D.
With a preface by Joyce Ann Schwarz

Keats Publishing ✚ New Canaan, Connecticut

Neither the author nor the publisher has authorized the
use of their names or the use of any of the material contained
in this book in connection with the sale, promotion
or advertising of any product or apparatus. Any such
use is strictly unauthorized and in violation of the rights
of Richard A. Passwater and Keats Publishing, Inc.

SELENIUM AS FOOD & MEDICINE

Pivot Original Edition published 1980

Copyright © 1980 by Richard A. Passwater

Printed in the United States of America

ISBN: 0-87983-237-1 (hardcover)
    0-87983-229-0 (paperback)

Library of Congress Catalog Number: 80-82325

PIVOT ORIGINAL HEALTH BOOKS are published by
Keats Publishing, Inc., 36 Grove Street (Box 876),
New Canaan, Connecticut 06840

TO THE MEMORY OF

Klaus Schwarz, M.D.
1914-1978

Contents

List of Illustrations

Preface by Joyce Ann Schwarz

Introduction by Julian E. Spallholz, Ph. D.

1   Introducing the selenium spectrum   3

2   Selenium against cancer   14

3   Selenium builds a healthy heart   46

4   Slowing down the aging process with selenium   63

5   Arthritis   84

6   Selenium strengthens your immune system   88

7   The growing danger of radiation   96

8   Cystic fibrosis   100

9   Muscular dystrophy   113

10   Cataract   118

11   Sexual function   123

12   Selenium detoxifies pollutants   128

13   The experts speak out   132

14   Recommended daily allowances   155

15   We need more selenium   160

16   Selenium supplements   183

17 Toxicity   197

18 Man does not live by selenium alone   207

19 Perspective   212

Appendix A Free radical theory   216

Appendix B Author's defense of protein
    missynthesis theory   225

Appendix C McCarty's hypothesis on
    macrophage   function   230

Glossary

Index

List of Illustrations

Fig. 1.1 Geographical distribution of selenium in the United States, 10

Fig. 1.2 Cancer incidence rate by state, continental United States (1959), 11

Fig. 2.1 Zero cancer extrapolations (breast, rectum, intestine), 17

Table 2.1 Relationship between selenium in crops and breast cancer rate, 19

Table 2.2 Cancer death rate in various areas, 19

Fig. 2.2 Cancer death rate versus selenium level (expressed as percentage difference from national average), 20

Fig. 2.3 Cancer death rate versus soil selenium level, 20

Table 2.3 Selenium concentration in human blood and human cancer death rate in various cities (1962-66), 22

Fig. 2.4 Relationship of selenium intake and breast cancer mortalities, 23

Fig. 2.5 Illustration of a free radical, 26

Fig. 2.6 Selenium is protective against carcinogen-produced cancer, 29

Fig. 2.7 Breast cancer reduced by selenium, 31

Fig. 2.8 Effect of selenium in preventing breast cancer, 32

Fig. 2.9 Properties and functions of selenium that may be related to its anticarcinogenic action, 37

Fig. 3.1 Heart degeneration due to selenium deficiency, 50

Table 3.1 World heart disease rates versus selenium intake, 54

Table 3.2 U.S. heart disease death rates and selenium, 55

Fig. 4.1 Abnormal cross-linkage of molecules caused by free radicals, 70

Fig. 4.2 Four steps in the free-radical attack on a cell, 72–73

Fig. 4.3 Metabolic pathways of the glutathione-Se enzyme complex, 74

Fig. 4.4 Accumulation of lipofuscin with age, 80

Fig. 4.5 Fluorometric quantification of lipofuscin in relationship to three different diets, 81

Fig. 6.1 Schematic showing the major components of the immune system, 91

Fig. 8.1 Organic changes in cystic fibrosis related to selenium deficiency, 102

Fig. 8.2 Maternal selenium levels related to congenital pancreatic cystic fibrosis, 107

Fig. 15.1 Maps of acid rain and deposition sensitivity in United States, 164–165

Fig. 15.2 Effects of selenium on mammalian life, 169

Table 15.1 Relative food levels of selenium, 171–176

Table 15.2 Selenium content in feedstuffs, 176–182

Table 18.1 Vitamin E and selenium deficiency diseases of animals, 210–211

Fig. A.1 Simplified schematic of major factors at work in the aging process, 224

Preface

My late husband, Klaus Schwarz, M.D., and his colleagues established the essentiality in human nutrition of three trace elements: selenium, chromium and fluorine. Klaus was also working with several other potentially important trace elements including tin, vanadium and nickel. I was proud of my husband's achievements when these discoveries were made nearly two decades ago, but only now are the benefits of this research beginning to be realized.

Scientists from five continents assembled in the spring of 1980 at the Second International Symposium on Selenium in Biology and Medicine. They discussed selenium's role in the prevention of such diseases as congestive heart failure, cancer and cataracts. The facts about selenium are well known in this small and exclusive scientific fraternity. The general public, however, knows very little about selenium. I am delighted, for this reason, that Dr. Richard Passwater has written this excellent book. It explains, simply and clearly, how adequate amounts of selenium may help to prevent and cure a wide range of diseases that now ravage millions of people. Dr. Passwater's statements and conclusions are based upon hundreds of scientific documents and clinical experiments.

Klaus told me often that his dream was to eradicate at least one disease through his scientific research. This desire, this dream, has already come true as Dr. Passwater points out in Chapter Three in which he explains how congestive heart failure was both prevented and cured in Mainland China through selenium supplementation. Dr. Passwater discusses in Chapter Two selenium's preventive effect against cancer, and in Chapter Twelve he tells how selenium can help us cope with pollution. Dr. Passwater explains in Chapter Ten how selenium may help to arrest and even reverse cataracts.

Rarely has a book been written with a greater potential for improving the health and welfare of people around the globe; it is for scientists such as my husband to discover great truths and it is for scientists such as Dr. Passwater to disseminate these truths.

I am indeed grateful to Dr. Passwater for dedicating this book to Klaus, and I am gratified that this extremely important information will now become available to millions of people. I am also grateful that, through Dr. Passwater, the dreams and accomplishments of Klaus Schwarz, M.D.—scientist, husband and father—will be better known and appreciated ... that Klaus's brief sojourn on this Earth was by no means transitory but will live on in better health for all Mankind.

Joyce Ann Schwarz

Introduction

Research in nutrition has over the years contributed much to our understanding of health and disease. As a scientific discipline, nutrition has only come of age during this century, having evolved from physiology and chemistry. Recognition of the diet in preventing disease, however, has been known since the British physician Lind published his *Treatise on Scurvy* in 1753. Later, in 1890, the Dutch physician Eijkman produced beriberi experimentally in chickens fed polished rice. Not until 1912 was the theory of disease and deficiencies of diets formalized when Casimir Funk wrote *The Etiology of the Deficiency Diseases*. The vitamine (sic) theory fully came of age in 1913 when McCollum and Davis observed that rats failed to grow when fed a refined diet. A fat-soluble substance extracted from butter and eggs and added to the refined diet restored growth (we now know this factor was vitamin A) to rats.

All of the vitamins we know today, A, B-complex, C, D, E and K were isolated and identified from food sources between 1926 and 1948. Most of these vitamins and the essential amino acids of protein were identified between 1930 and 1939—a golden decade of scientific achievement in this century.

Whereas the first half of this century was spent identifying vitamins, the essential fatty and amino acids, the latter half of this century to date has seen our understanding of our needs for the nutrients we call minerals increase dramatically. Many of the minerals we require are needed in our diets in quantities smaller than some vitamins. Until this century only iron and iodine were known to prevent dietary deficiency diseases. Between 1928 and 1935, copper, manganese, zinc and cobalt were recognized as necessary nutrients. Recognition of the need for molybdenum came in 1953. In 1957, ending more than a decade of almost continuous research, German-born physician Klaus Schwarz, then of the National Institutes of Health, identified selenium as a component of an organic constituent he called Factor 3 which prevented the nutritional deficiency disease dietary liver necrosis in rats. Since 1957, eight other minerals have been implicated to have nutritional value. None of these minerals, however, has generated such scientific and general nutritional interest as has selenium.

Selenium was named for the goddess of the moon, Seléne; but for nutrition it has been a shining star of the last decade. Not since the recognition of the mineral cobalt as part of vitamin B-12 in the mid 1950s has a mineral been recognized as having a precisely definable function in human or animal nutrition as has happened with selenium. In 1973, Dr. J. T. Rotruck (then at the University of Wisconsin) and his associates identified selenium as a necessary component of an enzyme, glutathione peroxidase.

As a component of glutathione peroxidase, and perhaps with other functions, selenium is believed by many people to protect cell membranes, prevent cardiovascular diseases, reduce the incidence of cancer, suppress arthritis, reduce aging and contribute generally to better health. It is probable that the formal recognition of selenium as a nutrient by the Food and Nutrition Board of the National

Research Council (1980), their issuance of dietary guidelines for selenium, recognition of Keshan Disease in the Peoples' Republic of China as a selenium deficiency disease, deficiencies of selenium in the diets of the Scandinavian countries and elsewhere may soon result in a significant portion of the world's population receiving selenium supplements. Animals already do.

Selenium as Food & Medicine by Dr. Richard A. Passwater is a timely, comprehensive compendium about the nutritional needs of people for selenium. It balances and places in perspective the scientific facts about selenium and our expectations for better health through its proper use, serving the public's need to know. *Selenium as Food & Medicine* is extensively referenced and up-to-date. It contains pertinent information about selenium, e.g. Keshan Disease, presented by research scientists from the United States and eleven foreign nations at the Second International Symposium on Selenium in Biology and Medicine held at Texas Tech University at Lubbock Texas in May 1980.

Julian E. Spallholz, Ph.D.
Lubbock, Texas

Selenium
as Food
& Medicine

ONE

Introducing the selenium spectrum

ITS time to stop ignoring the facts! If you want to maintain your good health, increase your resistance to disease, and assure a long and energetic life, it is vitally important that you increase your intake of selenium. Unfortunately, most people today are not getting enough of this essential mineral. The amount of selenium that is available in the soil to be made part of our food supply has decreased alarmingly. Modern fertilization practices, combined with the spread of acid rains, have reduced plant intake of selenium even in those areas where the soil contains enough of this vital element. In many instances it has become virtually impossible to obtain optimum amounts of selenium—even when the proper foods are consumed. Unfortunately, far too many diets already fall well below the National Research Council's recommended intake, mostly because of poor food choices, processing and over-cooking.

The purpose of this book is to report on the population studies, clinical trials and laboratory experiments that establish without doubt selenium's vital role in the prevention of heart disease and many forms of cancer. Proper use of selenium can also slow the aging process, strengthen the body's immune system and improve energy levels. There is, moreover, increasing evidence that selenium can help prevent and relieve arthritis, forestall the onset of cataracts, and improve resistance to infectious diseases. There are even strong suggestions that selenium can improve sexual health and possibly sexual function as well.

When I first joined the small group of researchers at work on selenium in the 1960s, my colleagues in the medical profession asked me pointblank: "Who needs it?" Now, only a few years later, selenium's vital role in the prevention of cancer and heart disease, along with numerous other beneficial effects, is well known among biochemists.

In 1979 Herbert H. Boynton, long associated with research in the selenium area, admitted: "It sometimes gets embarrassing to write about the multifarious and seemingly disparate biological roles of selenium. It begins to sound like some turn-of-the-century snake oil rather than what it is: an extremely important essential non-metallic trace element that functions effectively in the microgram range. Explaining how selenium functions biologically may help to restore credibility in its broad-ranging effects. Since selenium functions at the cellular and subcellular level, it is not surprising that it beneficially affects pathologies that are seemingly unrelated."[1]

Boynton's credibility point is well taken. Nevertheless, selenium does protect the membranes of each of our body's 60 trillion cells. In so doing, it prevents the decay of cellular function and also helps to protect those individual working parts of the cells called *organelles* (little organs). While we researchers understand selenium's wide-ranging role, we sometimes forget that others may not appreciate the role selenium plays in protecting the body against a whole host of diseases. During the past twenty years, I have demonstrated selenium's effectiveness repeatedly in my own laboratory; I even hold patents for formulations that include selenium and are related to this positive effect, although none of the patents are for commercial products. Among other selenium researchers with whom I've been privileged to correspond are Drs. Klaus Schwarz, Milton Scott, Al Tappel, Douglas Frost, Orville Levander, R. Shamberger and Gerhard Schrauzer. From their brilliant findings, I know that selenium is protective against cancer, heart disease and premature aging.

In the interest of the reader's best health, I wish to pass along my understanding of how selenium works to prevent so many different ailments common to man. My evidence will be presented in two ways—through summaries, discussions of experiments and interpretive graphs given in the text, and with appendices containing theories and further information for those who wish it (thus we will not unnecessarily divert the reader's attention from the main thrust of the text).

Each chapter that follows will report in full the evidence to substantiate statements made concerning selenium and its beneficial effects. These include:

Cancer

Current evidence suggests that improved selenium nutrition can reduce cancer risk. In recent studies, moderate amounts of selenium were added to the normal diets of animals who had been treated with cancer-causing substances, or who had evidenced a high, spontaneous incidence of cancer. In almost every case, *selenium supplementation substantially reduced the incidence of cancer*.

One particularly striking study by Dr. Gerhard Schrauzer of the University of California at San Diego included a control group of mice of a type that normally develops breast cancer spontaneously. Among the group that did not receive extra selenium, 83 percent developed breast cancer, while those mice whose diets were supplemented by selenium throughout their lives lived longer and developed cancer in only 10 percent of the cases. This finding is supported by the fact that in selenium-poor New Zealand the incidence of intestinal cancer in sheep dropped sharply after the sheep received selenium supplements.

At the human level, there is equally impressive evidence that increased selenium intake can protect people too. In those cases where regional nutritional selenium consumption has been effectively measured, a negative correlation between selenium intake and the incidence of many types of cancer—including breast, colon, ovary, pancreas, prostate, lung and bladder—has consistently resulted. In other words, the more selenium consumed, the lower the incidence of cancer.

Other evidence suggests that the blood selenium level of cancer patients is usually low or low-normal. Recently several physicians have found that when sufficient selenium is ingested by cancer patients to raise their blood levels of selenium to the desired range, their tumors began to shrink (see Chapter Two). This is not to say that selenium cures cancer, but at this stage of research it does imply that a selenium deficiency can interfere with a cancer cure.

Heart disease

Studies conducted by such researchers in cardiovascular diseases as Dr. Raymond Shamberger of the Cleveland Clinic and Dr. Johan Bjorksten of the Bjorksten Research Foundation suggest that selenium is an important protective factor in high blood pressure, stroke, heart attack and hypertensive kidney damage. Inhabitants of areas low in selenium have high rates of heart disease, while those with high selenium intake experience low rates of heart disease.

This protective role of selenium may be the result of several different mechanisms. For one, selenium is necessary for the health of the heart muscle itself. It can, for example, improve the function of *mitochondria*—the energy-producing units of cells—by protecting them from lack of oxygen. This may account for the fact that selenium supplementation is effective in the treatment of chest pains associated with heart disease (angina pectoris). Selenium is also required for the production of a hormone-like substance called *prostaglandin*, which helps regulate blood pressure. And selenium detoxifies cadmium, a pollutant which produces high blood pressure.

Chinese physicians recently reported that a congestive heart disease—one that affected children especially—was prevalent throughout vast areas of rural China. Since this land is known to be low in selenium, the physicians set up a carefully controlled study. In one commune, the people were allowed to eat their regular diets, but they were also given 1000 micrograms of selenium each week. Before long, the incidence of this heart problem—called Keshan disease in China—dropped to zero, and children who were already suffering from heart trouble became well. Meanwhile, in a neighboring commune, the children continued to experience a high rate of heart disease. When selenium was added to their diet as well, no new cases developed and those already afflicted improved markedly.

Scarcity

The distribution of selenium in the soil varies greatly throughout the world, including the United States. Small percentages of this vital substance are present in such heavily populated areas as New England, the rest of the East Coast, and the Great Lakes area, as well as the Northwest (see map, Figure 1.1). For the most part, areas which have been glaciated in the prehistoric past tend to have extremely low selenium levels.

But soil content is only part of the problem. Even in areas where selenium is present in the soil, it is being made even scarcer because of modern fertilization practices and acid rain. The level of sulfur, which inhibits the absorption of selenium by plants, is being raised by the use of artificial fertilizers that are rich in sulfur. Moreover, heavy deposits of a sulfate in the atmosphere resulting from the burning of coal and oil are further limiting the amount of selenium available from the soil. For example, selenium deficiency became a significant problem for the sheep industry in Oregon only after high sulfur fertilizers were introduced.

The selenium content of plants tends to reflect the availability of soil selenium. Plants do not require selenium for growth and therefore do not concentrate it. So-called "experts" who assure you that plants have normal amounts of minerals or they would not grow do not know what they are talking about. Plants need only fourteen or so nutrients, while man needs fifty-five. We need many more minerals for health than plants do for growth. Wheat grown in the selenium-rich soils of the Midwest contains up to 100 parts per million (ppm) of selenium, while that grown in the low-selenium state of Washington contains less than 0.1 ppm. Those who rely on plant foods grown in low-selenium or high-sulfur regions obviously run a strong risk of developing selenium deficiencies, just as foods grown in the goiter belt produced iodine deficiency before iodized salt was made available.

The National Research Council has now established a recommended daily intake of 50 to 200 micrograms of selenium. But many so-called "good" diets average only 35 to 60 micrograms per day. This means that many people who are subsisting on poorly designed diets receive far less than the recommended amount of selenium. Indeed, several eminent scientists have stated that literally millions of Americans are getting less than the recommended amounts of selenium. Drs. Schrauzer and Shamberger agree that a daily adult selenium intake should equal 250 to 350 micrograms. I believe that some people should eat more—but not exceeding 700 to 800 micrograms for long periods unless under the close supervision of a physician.

FIG. 1.1

Geographical distribution of selenium in the United States

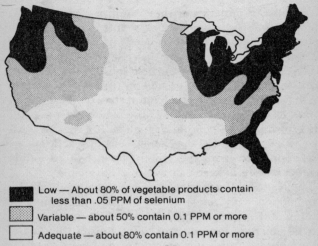

Low — About 80% of vegetable products contain less than .05 PPM of selenium

Variable — about 50% contain 0.1 PPM or more

Adequate — about 80% contain 0.1 PPM or more

Source: Data from USDA Technical Bulletin No. 758 (1967)

The need for supplementation

Some will argue that there is no need to supplement a well-balanced diet. I will present evidence in this book that has led to my conviction that we can all benefit from selenium supplementation, and that to achieve optimum health some kind of supplementation program is mandatory. Those old-line nutritionists who still refuse to acknowledge the many shades of differences that exist between illness, absence from illness, average health and

FIG. 1.2
Cancer incidence rate by state, continental United States (1959)

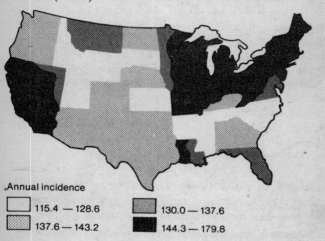

Annual incidence

☐ 115.4 — 128.6	▨ 130.0 — 137.6
▦ 137.6 — 143.2	■ 144.3 — 179.8

Source: Schrauzer, G. 1959. *Bioinorganic Chemistry* 2(4)

optimum health will howl in protest over my findings. These same people, I predict, are those who have never conducted selenium experiments or have never measured the selenium content of typical diets at the consumption stage (as opposed to food table estimates, which are not representative). I have done both.

The same nutritional studies that led to iodized salt and enriched flour (a misnomer that represents only the *partial* replacement of a few of the nutrients lost during the milling of whole wheat into white flour) apply to selenium. Many Americans simply do not have enough selenium in their diet to prevent health problems. No matter how well food is selected, if there isn't any selenium in the soil there can't be any in the crops or livestock. Bear in mind, though, that I am *not* suggesting mandatory supplementation. I would be strongly opposed to adding selenium to everyone's drinking water, for example. But everyone should have a choice between buying selenium-enriched salt or flour, just as we have with iodized or noniodized salt. Without that choice, we must all depend on restoring selenium to the soils in the form of fertilizers or by using personal food supplements. Ideally, selenium should be of natural origin rather than in the form of inorganic salts.

In the pages that follow, I will attempt to explain and confirm these important facts about selenium, many of which have been introduced in this opening chapter:

☐ Selenium's relationship to human health is an established fact.

☐ The recommended daily intake of selenium is 50 to 200 micrograms.

☐ Average American diets may provide only 35 to 60 micrograms in some areas—and much less in poorly designed diets.

☐ Several highly regarded scientists have stated that millions of Americans receive less than optimum amounts of selenium in their diets.

☐ Many people live in areas with low soil selenium availability.

☐ Selenium can help prevent many forms of cancer.

☐ Selenium can help protect against heart disease.

☐ Selenium strengthens your immune system.

☐ Selenium may improve your energy level.

13 Introducing the selenium spectrum

☐ Selenium helps prevent or relieve arthritis.

☐ Selenium can slow down evidences of aging and help make you look younger.

☐ Selenium detoxifies several heavy metal pollutants including cadmium, mercury and probably lead.

☐ Selenium may prevent the onset of cataracts.

☐ Selenium may affect fertility, sex drive and human reproduction.

REFERENCES

1 Boynton, Herbert H. January 1979. *The American Chiropractor* no. 52.

TWO

Selenium against cancer

I F there is no valid scientific reason to doubt selenium's effectiveness as a protective against many types of cancer, the question remains: will it protect against *all* types of cancer? It may be that *no* cancer cure will work in patients with selenium deficiencies. It has been found recently that cancer patients will suddenly respond to conventional therapy—that is, their tumors start to shrink—once their blood levels of selenium have been brought up to normal with supplements.

It is time to study cancer prevention in terms of selenium deficiencies. Referring to a conference at the International Association of Bioinorganic Scientists the

San Diego *Union* reported on January 10, 1979: "Some of the nation's leading nutritional scientists meeting here said they believe there is sufficient evidence that selenium can reduce some cancers to conduct a national trial of the trace element on humans ... A majority of the scientists said they are satisfied that available data is sufficient to indicate that supplementation of the diet with 100 to 200 micrograms of selenium will reduce the occurrence of some cancers in humans ... They said in a poll that they favor a human field trial to test the effectiveness of dietary selenium against cancer and suggested the study involve 10,000 to 15,000 persons taking selenium and an equal number taking a placebo."[1]

Dr. Gerhard Schrauzer of the University of California, San Diego, one of the chairmen at this conference, had reported a few weeks earlier to the Workshop on Chemoprevention of Cancer at the National Cancer Institute in Bethesda, Maryland, that "the key to cancer prevention lies in assuring the adequate intake of selenium, as well as other essential trace elements."[2]

Dr. Schrauzer also commented in an article published in *Family Circle* magazine (June 1978) on the role of selenium in cancer patients: "If a breast cancer patient has low selenium levels in her blood, her tendency to develop metastases (other tumors) is increased, her possibility for survival is diminished, and her prognosis in general is poorer than if she has normal levels."[3]

"If every woman in American started taking selenium today or had a high-selenium diet," Dr. Schrauzer added, "within a few years the breast cancer rate in this country would drastically decline."[4]

At the same time, Dr. Schrauzer pointed out that "selenium is a giant step toward preventing cancer—it is a major breakthrough. It is one of the most efficient agents in stimulating the natural defense mechanisms against cancer. If selenium were used properly as a preventive measure against cancer, I think it's possible that it would enable us to cut the mortality rate from almost all cancer by 80 to 90 percent in this country."[4]

He was referring to the evidence that selenium helps prevent ten of the most common forms of cancer that in 1978 caused approximately 250,000 deaths. According to his statement then, 200,000 to 225,000 lives could be saved annually. His figures indicate that in the U.S. the projected zero cancer mortality occurs at twice the average blood level of selenium. This level could be reached with an appropriate diet.[5] (See Figure 2.1.)

Dr. Schrauzer has conducted extensive research on selenium and cancer since the mid-1960s. In 1971 he reported that one of the blood tests for cancer was actually a test for blood selenium level. He reasoned that low levels of selenium were associated with greater susceptibility to cancer.[6,7]

Other cancer researchers have expressed similar comments on the preventive role of selenium. Dr. Pietro Gullino, chief of the Laboratory of Pathophysiology and chairman of the Breast Cancer Task Force, adds: "There is no question that there is a certain relationship between areas poor in selenium and areas with a high frequency of cancer in general. It is clear at this point selenium is considered an important element in the whole story of cancer in general."[4]

Dr. Douglas Frost, former researcher at Dartmouth Medical School's Trace Element Laboratory, puts it rather succinctly: "There is damn good evidence that selenium has anti-cancer value. If we want to avoid getting cancer, we should be sure to get enough selenium."[4]

Dr. Charles Shaw of the M.D. Anderson Hospital

FIG. 2.1   Zero cancer extrapolations

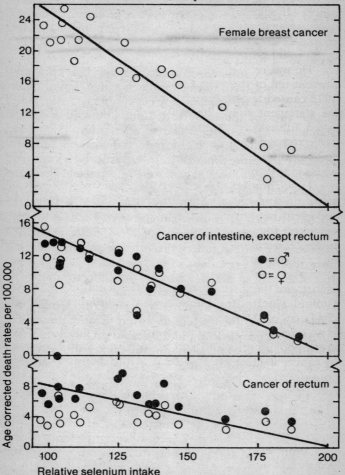

Female breast cancer

Cancer of intestine, except rectum

● = ♂
○ = ♀

Cancer of rectum

Age corrected death rates per 100,000

Relative selenium intake

Source: Schrauzer, White, and Schneider, 1977. *Bioinorganic Chem.* 7:37.

and Tumor Institute in Houston, Texas, who has reduced cancer in laboratory animals, states with scientific conservatism: "There seems to be a causal relationship between having high-selenium levels and not having cancer. It all fits a common picture."[4]

These scientific observations are based on epidemiological studies, laboratory experiments and clinical observations. Based on epidemiological studies, Dr. Raymond Shamberger of the Cleveland Clinic Foundation advises people to increase their intake of selenium to 200 micrograms a day, because "it can reduce the cancer rate dramatically for some types of cancer, particularly cancer of the colon, breast, esophagus, tongue, stomach, intestine, rectum, and bladder."[4]

It was Dr. Shamberger who discovered in 1969 that the blood levels of cancer patients were low in selenium. His laboratory experiments in 1966 had, in fact, indicated the presence of selenium deficiencies in cancer. While the normal level is more than 18 micrograms of selenium per 100 milliliters of blood, many of the cancer patients he studied had only 12 to 15 micrograms.

Dr. Shamberger, together with Dr. Frost, studied the relationship between selenium in crops and the cancer incidence in people living in those areas. They found that the lower the level of selenium, the higher the incidence of cancer. The results are clearly revealed in Tables 2.1 and 2.2, and in Figures 2.2 and 2.3.[8]

In another study, Drs. Shamberger and C. Willis found that healthy persons between the ages of fifty and seventy-one averaged 21.7 micrograms of selenium per 100 milliliters of blood, whereas cancer patients of the same age range averaged only 16.2 microliters per 100 milliliters. The worst cancer cases had the lowest selenium levels (13.7, 13.9, and 14.3).[9]

TABLE 2.1
Relationship between selenium in crops and breast cancer rate

Selenium crop concentration ppm	Breast cancer rate compared to nutritional average
0.03	11% above average
0.05	9% above average
0.10	20% below average
0.26	20% below average

Source: Shamberger, R. and Frost, D. 1969. *Canadian Medical Association* 100:682.

TABLE 2.2
Cancer death rates in various areas

Selenium level	(ppm) (soil)	No. states	Cancer death rate (per 100,000)
Very High	above 0.26	6	392
High	0.10 to 0.25	19	430
Medium	0.06 to 0.09	11	450
Low	0.01 to 0.05	20	516

The 1968 55-64 age-specific cancer death rate for the white males in states located in regions of selenium occurrence.

Source: Shamberger, R. May 11–13, 1976. From the *Proceedings of the symposium on selenium-tellurium in the environment.* Univ. of Notre Dame, p. 262.

FIG. 2.2 ,2.3
Cancer death rate versus selenium level

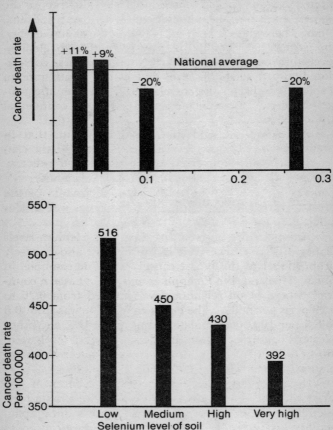

Sources: Passwater drawings based on data from Shamberger, R. and Frost, D. 1969. *Canadian Medical Association Journal* 100:682.

Other epidemiological data reveal that Rapid City, South Dakota, has the lowest cancer rate of any city in the United States. The citizens of that city also report the highest measured blood selenium levels. But in Lima, Ohio, where the cancer rate is twice that of Rapid City, the citizens have only 60 percent of the blood selenium levels disclosed in Rapid City.[10] Note the inverse correlation between selenium level and cancer death rate in Table 2.3. indicates the correlation of soil selenium concentration to cancer incidence rates for the continental United States.

Dr. Schrauzer and his colleagues found that in twenty-seven countries surveyed (see Figure 2.4.) the cancer death rate was inversely proportional to the dietary intake of selenium in the typical diets of those countries.[11-13] Cancers involved in the study included tumors of the breast, ovary, colon, rectum and prostate, as well as leukemia.

Another study showed that the blood selenium levels of persons from seventeen countries were also inversely proportional to the breast cancer rate. For example, in Japan, Thailand, the Philippines and other Eastern countries, where blood selenium levels ranged from 0.26 to 0.29 parts per million, the breast cancer rate was from 0.8 to 8.5 per 100,000 population. But in the U.S. and other Western countries, where the blood selenium levels were lower (0.07 to 0.20 ppm), the breast cancer rate was higher (16.9 to 23.3 per 100,000).

In Venezuela, the death rate from cancer of the large intestine is 3.06 per 100,000. Venezuela has a high selenium content in its soils, while ours is low. Japan, another high-selenium country, has less breast cancer, as we have noted, and it also enjoys a lower lung cancer death rate—12.64 per 100,000 compared to our rate of

TABLE 2.3

Selenium concentration in human blood and human cancer death rate in various cities (1962–66)

City	Blood Se (mcg/100 ml)	Cancer deaths per 100,000 pop.
Rapid City, S.D.	25.6	94.0
Cheyenne, Wyo.	23.4	104.0
Spokane, Wash.	23.0	179.0
Fargo, N.D.	21.7	142.0
Little Rock, Ark.	20.1	176.0
Phoenix, Ariz.	19.7	126.7
Meridian, Miss.	19.4	125.0
Missoula, Mont.	19.4	174.0
El Paso, Tex.	19.2	119.0
Jacksonville, Fla.	18.8	199.0
Red Bluff, Calif. (Tehama Co.)	18.2	176.0
Geneva, N.Y.	18.2	172.0
Billings, Mont.	18.0	138.0
Montpelier, Vt. (Wash. Co.)	18.0	164.0
Lubbock, Tex.	17.8	115.0
Lafayette, La.	17.6	145.0
Canandaigua, N.Y. (Ontario Co.)	17.6	168.0
Muncie, Ind.	15.8	169.0
Lima, Ohio	15.7	188.0

Source: Shamberger, R. and Willis, C. June 1971. CRC Reviews.

FIG. 2.4
Relationship of selenium intake and breast cancer mortalities

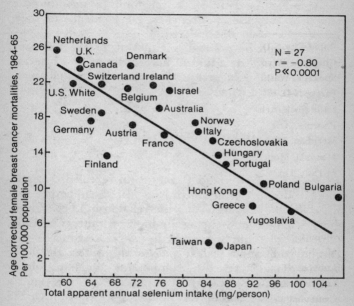

Source: Schrauzer, G., White, D. and Schneider, C. 1977.
Bioinorganic Chemistry, vol. 7, p. 36.

36.86 per 100,000. This same inverse relationship was found in a 1971 study to correlate with the level of selenium in local cow's milk.[14] And finally, in a 1972 study made in New Zealand, where selenium was introduced to sheep in anthelmintic drenches, the incidence of sheep cancer practically disappeared.[15]

It should be pointed out that all of the foregoing epidemiological data do not necessarily prove cause and effect—they only suggest possible relationships that should be researched further.

Laboratory experiments

In 1949 C. Clayton and L. Baumann revealed that selenium reduced cancer one-third.[16] Dr. Shamberger and his colleagues began testing the relationship between dietary selenium and chemically induced cancers in 1966. One line of thought current in cancer research is that certain compounds called *antioxidants* protect the body against damage caused by compounds containing highly reactive fragments called "free radicals." At this point it seems fitting to define free radicals since they are at the heart of cancer research.

Normal cells stop growing when they meet neighboring cells, but cancer cells keep on growing uncontrollably; they no longer respect boundaries. This type of cell membrane damage can arise from reactive fragments of molecules (free radicals).

Atoms and molecules are the most stable chemical forms since they have even numbers of positive-charged protons and negative-charged electrons. Thus atoms and molecules have no electrical charge. If an atom or molecule either loses or gains an electron, then the atom or molecule becomes "charged."

The charged atom or molecule is called an "ion." Positive-charged ions are called "cations," while negative-charged ions are called "anions." Since opposite charges attract each other, cations and anions are attracted to each other with the resultant formation of a neutral molecule.

Many compounds readily dissociate into their ions; such an example would be sodium chloride (table salt) forming sodium and chloride ions when dissolved in water. Other compounds are difficult to ionize.

A molecule normally can be ruptured to produce ions. The normal manner in which the attractive forces (bonds) are ruptured results in even pairs of electrons existing in both ions. In the chemist's jargon, this normal process is described as a two-electron chemical bond cleaving asymmetrically.

A second type of molecular rupturing can also occur, although this process is comparatively rare. The molecule can rupture, leaving the two fragments each having an uneven number of electrons. These fragments are called "free radicals." A free radical is defined as an atom or group of atoms with an unpaired electron (see Figure 2.5).[17] Free radicals normally are neutral, but can be charged. As an example, the superoxide radical is a negative radical ion.

The unpaired electron gives the free radical unbalanced magnetic energy with the result that free radicals will stabilize themselves by magnetically attracting a neighboring electron from any other nearby molecule. This is a "free-radical reaction." Ions only interact with other ions of opposite charge. Free radicals will react with anything.

Free radicals are produced by nuclear and x-radiation,

FIG. 2.5
Illustration of a free radical

SUPEROXIDE (O$_2^-$)
(free radical)

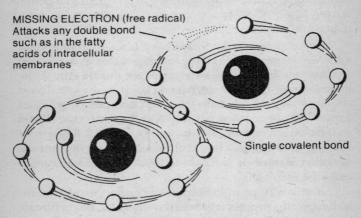

MISSING ELECTRON (free radical)
Attacks any double bond
such as in the fatty
acids of intracellular
membranes

Single covalent bond

The superoxide radical is used to illustrate the structural nature of a free radical. In this particular instance, there is a single covalent bond and a missing electron.

Source: Halstead, B. 1979. *The Scientific Basis of EDTA Chelation Therapy.* Colton, California: Golden Quill Publishers, p. 53. Artist: R. H. Knabenbauer

sunlight, metals and metabolism. A few enzymatic reactions work via free radicals in the body, but generally speaking most free radicals produce deleterious effects.

The damage done by free radicals is very extensive because once a free radical is formed, it usually triggers a chain reaction that can generate severe damage before the free radicals are destroyed. One free radical can produce chain reactions altering thousands and thousands of molecules.

Most free radicals persist for only a few thousandths of a second, but stable free radicals such as triphenylmethyl and 2,2-diphenyl-1-picrylhydrazyl can be stored in a bottle. A free radical can be detected by its paramagnetic resonance and phosphorescence due to its unpaired electron.

Free radicals can be formed as a reaction to unneeded superoxidation in the cell (peroxidation), or as a reaction to insufficient oxygen. Nobel Laureate (1931) Otto Warburg long ago noted that normal cells use oxygen-based reactions as their source of energy, but cancer cells can form from cells not receiving adequate oxygen. These oxygen-deficient cells sometimes change to a glucose-based chemistry to derive their source of energy instead of dying. If the cells make this switch, they become cancer cells.

Antioxidants protect body cells against unwanted reactions with oxygen but allow the desirable oxygen reactions to proceed without interference. Many selenium compounds are antioxidants, or antiradicals.

In a series of experiments in which carcinogens (cancer-causing compounds) were applied to the skin of mice, dietary selenium had a strong protective effect. Selenium deficiency increased the number of mice that developed cancer as well as the number of tumors found in each mouse. Extra dietary selenium reduced both numbers in proportion to the amount of selenium provided.

Other experiments tested selenium's ability to protect animals against carcinogens consumed in the diet. Comparisons were made by Drs. Clayton and Baumann between normal diets and those fortified with selenium using azo dyes.[16] Similar tests were run by Dr. J.R. Harr and others with a carcinogen called FAA (1972),[18] and by myself with another carcinogen known as DMBA (1969–72).[19] Dr. Lee Wattenberg has conducted similar experiments with antioxidants other than selenium.[20]

My experiments were made with several antioxidants used in synergistic combination so as to provide animal protection at the lowest total dosage of antioxidants possible. The incidence of stomach cancer to be expected in mice given DMBA is 85 to 90 percent. By adding mixtures of water- and fat-soluble natural and synthetic antioxidants, including selenium, to their diet, that figure was reduced to 5 to 15 percent. In these experiments, the mice I used had been raised from weaning on the experimental diets, and they continued to receive the same diets after being fed one dose of the DMBA.

Dr. Harr and his colleagues fed their test animals a diet that included FAA on a constant basis during the entire experiment. Various amounts of selenium were added to the diet concomitantly with the carcinogen to four groups of twenty mice each. Group one received 150 ppm FAA and 2.5 ppm added selenium; group two received 150 ppm FFA and 0.5 ppm added selenium; group three received 150 ppm FAA and 0.1 ppm added selenium; and group four received 150 ppm FAA and *no* added selenium. After 210 days, eighty percent of groups three and four had developed cancer, compared to only ten percent in group two and three percent in group one. The selenium had an undeniably protective effect (see Figure 2.6).

Drs. Maryce Jacobs and Clark Griffin, while they were working at the University of Nebraska, demonstrated that selenium lowers the incidence of intestinal and rectal tumors in rats exposed to 1,2-dimethylhydrazine and methylazoxy-methanol acetate.[21] They also found that se-

FIG. 2.6

Selenium is protective against carcinogen-produced cancer

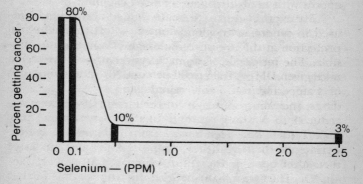

Source: Passwater drawing based on data from Harr, J., Exon, J., Whanger, P. and Weswig, P. 1972. *Clinical Toxicology* 5(2):187–94.

lenium lowers the incidence of liver cancer in rats caused by 3'-methyl-4'-dimethyl-aminoazo-benzene.[22]

In February 1978, at a conference on preventing cancer at the National Cancer Institute in Bethesda, Maryland, considerable emphasis was given to the role of selenium in preventing cancer. Dr. Clark Griffin, now at the M.D. Anderson Hospital and Tumor Institute in Houston, Texas, reported at the time that by adding selenium to drinking water or by feeding selenium in the form of high-selenium yeast, rats that were exposed to three different kinds of cancer-causing chemicals can be protected from colon and liver cancer. Dr. Griffin's group also revealed that selenium can prevent the conversion of potentially cancer-causing chemicals into other harmful forms. Experiments conducted by Dr. Charles R. Shaw, also of M.D. Anderson Hospital, showed that selenium reduced the bowel cancer rate from 87 percent to 40 percent in animals that had been fed carcinogens.

In 1974 Dr. Gerhard Schrauzer tested selenium against the natural cancer incidence in mice.[23] He was able to reduce the incidence of spontaneous breast cancer in susceptible female mice from 82 percent to 10 percent merely by adding traces of selenium to their drinking water. (See Figure 2.7 for related data.)

In addition to this eightfold reduction of cancer incidence, Dr. Schrauzer's study shows that even among the 10 percent of the selenium-supplemented mice that did develop cancer, the disease appeared 50 percent later than it did among the control animals. Moreover, the tumors were "less malignant" and the survival time of the selenium-supplemented animals was 50 percent longer. In a less cancer-prone strain of mice, the breast cancer may have been totally prevented.

Dr. P.D. Whanger of Oregon State University studied the effects of selenium on the formation of mammary tumors.[24] He and his colleagues fed groups of twenty to twenty-five mice one of the follow diets: lab chow, lab chow plus 0.5 ppm selenium in water, and lab chow plus 2.0 ppm selenium in water. The tumor incidence was 80, 25 and 25 percent, respectively (see Figure 2.8).

Dietary fats

We hear so much about fats in the diet causing cancer—everyone is warning us that we are eating too many fats. Fats are not necessarily the problem. The problem is that we are eating too many fats for the amount of antioxi-

FIG. 2.7

Breast cancer reduced by selenium

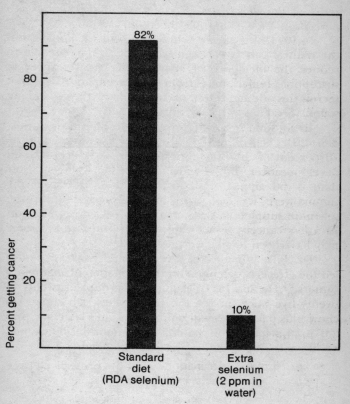

Source: Passwater drawing based on data from Schrauzer, G. and Ishmael, D. 1974. *Annals of Clinical Laboratory Science* 4:441–7.

FIG. 2.8

Effect of selenium in preventing breast cancer

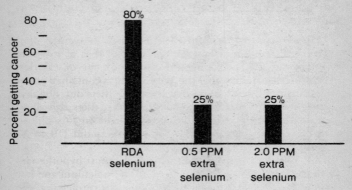

Source: Passwater drawing based on data from Whanger, P., Tinsley, I., Schmitz, J. and Exon, J. May 1980. Second International Symposium on Selenium in Biology and Medicine, Texas Tech University, Lubbock, Texas.

dants such as selenium that we get. To put it another way, *the real problem is the deficiency of selenium and other antioxidants*, a situation that is aggravated by the burden of extra dietary fat.

The May 3, 1976, issue of *Chemical and Engineering News* added more information about the protective effect of selenium against dietary fats.

The association between high selenium levels in the diet and a lower-than-average cancer rate was suggested in a paper delivered by Dr. Christine S. Wilson, a nutritionist at the University of California, San Francisco. She told the FASEB meeting that high selenium levels in the diet may explain why the breast cancer rate is substantially lower in Asian women than in women from western countries.

(Note: The rate of breast cancer in Asian countries is about one-seventh of the U.S. rate.)

After comparing the nutrient content of an average non-Western diet supplying 2500 calories to that of a typical American diet providing the same number of calories, Wilson determined that the Western diets contained about a fourth of the selenium that the Asian diets did. She says that it is also significant that the Asian diets contained much less 'easily oxidizable' polyunsaturated fats (7.5 to 8.7 grams a day) than the Western diets did (10 to 30 grams).

The University of California nutritionist hypothesizes that it is the dietary combination of high selenium and low polyunsaturated fatty acids that may be protecting the Asian women against breast cancer. She notes that selenium is a component of the glutathione peroxidase system. Because the enzyme acts to inhibit the oxidation of the unsaturated fats, it blocks the formation of peroxides and free radicals, both of which are believed to trigger various forms of cancers.

The concept that polyunsaturated fats are more harmful than saturated fats has not caught on yet.

Free radicals are produced by several processes in the body. The healthy body normally can neutralize the self-produced free radicals before an excessive rate of damage occurs. However, polyunsaturated fatty acids form highly reactive self-propagating free radicals when initiated by a "starter" free radical. Whenever the intake of polyunsaturated fatty acids greatly exceeds that necessary for function, the risk of lipid peroxidation increases, as does the probability of cancer unless protected by additional antioxidants.

Today's trend toward increasing the ratio of polyunsaturated fats to saturated fats needs close examination if this is true. My writings on cholesterol and heart disease voice this concern along with my concern for unbalancing normal diets in trying to avoid dietary cholesterol. Cholesterolophobia, the needless fear of cholesterol by healthy people, has caused many people to increase their polyunsaturated fat intake far beyond that advised by physicians; the "average" American today ingests about two to three times that of Americans twenty years ago. A large number of studies alarmingly show proportionate increases in cancer (especially mammary) with increasing dietary polyunsaturate levels.

Pearce and Dayton have reported that humans fed four times the polyunsaturates of controls developed a significantly higher number of cancers: thirty-one of 174 as opposed to seventeen of 178.

This is not to say polyunsaturates cause cancer. It is to emphasize the need for a balanced diet and a balance of antioxidants and polyunsaturates. Nor is it to say that polyunsaturates are harmful, or that they should necessarily be reduced if one's diet is unbalanced. It is important to have adequate polyunsaturates for proper nutrition.

But extreme measures to drastically change the polyunsaturate/saturate ratio, especially with unbalanced antioxidants, appear to be a great risk, according to available information.

Typical animal experiments show that DMBA-induced cancer rates increase with polyunsaturates. Shamberger found that adding corn oil to the DMBA carcinogen promoter, croton resin, greatly increased the number of cancers.[25]

Scientists have been brainwashed that saturated fats were evil because they were alleged to cause heart disease and that polyunsaturates were good because they lowered blood cholesterol levels. This line of reasoning did not stand up to the experimental data developed in the 1970s, but old concepts die slowly. (For details, see Chapters One to Nine of *Supernutrition for Healthy Hearts*, Passwater, R., New York: Dial Press, 1977, or Jove, 1978.)

An indication that the medical community is finally coming around to the new evidence can be seen in the following International Medical News Service story released in August, 1979:

'Vegetable fats may be as important as animal fats in causing colorectal cancer,' Morris S. Zedeck, Ph.D., said at the International Symposium on Colorectal Cancer.

'The action of both animal and vegetable fats is similar; both increase the amount of bile acid excreted. Many researchers believe this is a major contributing factor in colorectal cancer,' said Dr. Zedeck, of the Memorial Sloan-Kettering Cancer Center, New York.

'Get rid of the idea that animal fat alone is the cause of colon cancer. It's not true,' he said.

Many different elements may go into the protection of the body from colorectal cancer. *One may be the trace element selenium.*

'Epidemiologic studies indicate there is a decrease incidence of colon cancer in those geographic areas having high levels of selenium in the soil. Blood selenium levels in patients with gastrointestinal cancer were lower than in the normal population,' Dr. Zedeck said.

'Animal studies are even more persuasive. Selenium has been found to decrease the incidence of both spontaneous tumors and chemically induced tumors. Although researchers are still struggling to determine the mechanism, selenium has also been shown to have a beneficial effect on hepatic, mammary and colon tumors,' he said.

Other elements are being studied to see what effects, if any, they have on colorectal cancer. These include dietary fats, genetics, bile acids, fiber and vitamin A.

'A fat-enriched diet without the lipotropic substances choline and methionine or diets deficient in vitamin A can augment the induction of colon tumors by chemical carcinogens,' Dr. Zedeck said.

In 1975, Dr. Birger Jansson of the M.D. Anderson Hospital in Houston presented studies showing the high incidences of colon cancer in low selenium areas.[26]

It is interesting to note that Puerto Rico has only 30 to 40 percent of the colon cancer and breast cancer rate as the U.S. Its people eat an average of 106 grams of animal fat daily, which represents 88 percent of their total fat intake. Americans have a lower percent of saturated fat (62 percent) and more cancer.

Residents of the Netherlands and Finland eat 100 grams of animal fat daily on the average. The cancer rate is twice as high in the Netherlands, possibly because more polyunsaturated fat is consumed there. In Finland, 88 percent of the daily fat intake is saturated, while in the Netherlands only 65 percent of the daily fat intake is saturated. Dr. Jansson and colleagues have demonstrated that selenium inhibits cancer by means of human epidemiological studies, experimental animal studies, mutagenesis assays and assays with human lymphocytes in culture. In one animal test with carcinogens, they found that selenium reduced the number of tumors induced by DMH by three-fold, and by MAM by nearly one half.

We must dismiss the concept that more polyunsaturates are necessarily good. All fat in excess is harmful, and the new evidence implies that of the two, excess saturated fats are less harmful than excess polyunsaturated fats. When kept to about 30 percent of the daily diet, it makes little difference how much of each fat type is included, as long as there is at least 10 percent of both.

Mechanisms

Selenium is involved in several mechanisms that protect the body from cancer (see Figure 2.9). It is involved at the first line of defense at the cell membrane as part of the antioxidant enzyme glutathione peroxidase. This protects against free radicals which will be discussed in further detail in Chapter Four.

FIG. 2.9

Properties and functions of selenium that may be related to its anticarcinogenic action

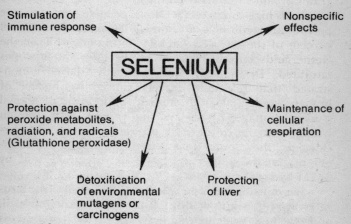

Source: Schrauzer, G. 1978. *Inorganic and Nutritional Aspects of Cancer.* New York: Plenum Press, p. 330.

In my opinion, the primary role of selenium is that glutathione peroxidase breaks down the epoxide form from the reaction of carcinogens with aryl hydrocarbon hydroxylase. The compounds that we call carcinogens are only the parent compounds or "pro-carcinogens." The "active" carcinogens are the epoxides formed within our bodies.

The latest information available further confirms the anti-cancer effect of selenium. The three abstracts reprinted below are taken from papers presented at the second International Symposium on Selenium in Biology and Medicine held at Texas Tech University in Lubbock, Texas, on May 13–16, 1980. At this symposium, Drs. Clark Griffin and Helen Lane of the University of Texas Health Science Center confirmed earlier findings that selenium reduces the incidence of cancer in animals that have been fed cancer-causing agents. Their abstract follows:

> Increasing numbers of chemicals reportedly inhibit or delay the formation of tumors in animals exposed to a variety to cancer-causing agents. These preventive chemicals include the retinoids, butylated hydroxytoluene or anisole, disulfiram, ethoxyquin, protease inhibitors, some prostaglandins, cyclic nucleotides, selenium compounds and others. Several studies related to the inhibition of chemically induced cancers by administration of selenium compounds have been conducted in our laboratories. Albino rats injected subcutaneously with dimethylhydrazine develop a near 100 percent of adenocarcinomas of the colon. However, the concurrent administration of *four to five parts per million selenium (as Na_2SeO_3)* in the drinking water or in the form of a *Se enriched yeast* added to the diet resulted in a greater than 50 percent reduction in the number of colon tumors at the end of the study. An extension of this same approach also provided indication that Se inhibited the induction of tumors in rats given the

active hepatocarcinogens, 2-acetylaminofluorene or 3' methyl-4-dimethylaminoazobenezene. Early findings suggest that the mechanisms of this selenium inhibition or prevention of carcinogenesis may involve the metabolism of carcinogens, DNA binding and repair and other parameters that will be discussed. A most important aspect is the possible projection of this approach to prevention of cancer in humans. Selenium levels and glutathione peroxidase activities were evaluated in a human population as a prerequisite to further studies of selenium as a chemopreventative agent. Glutathione peroxidase activities and selenium concentrations were determined in erythrocytes and plasma from two hundred healthy subjects. These selenium parameters were also determined from patients, ages fifty to seventy years, who had untreated squamous cell carcinoma in the head and neck region. The selenium parameters were compared between ages and sexes. The patients with cancer had slightly lower values for the selenium parameters than the healthy population. There was a slight trend for erythrocyte glutathione peroxidase activities and selenium concentrations in the healthy population to decrease with age but no difference was found between sexes. Explanation for the differences between the healthy and cancer populations will be discussed.

At the same symposium, Dr. Milner of Indiana University reported that selenium inhibits the growth of transplanted tumors. Dr. Milner's abstract follows:

Recent studies in our laboratory have shown that selenium is capable of retarding the growth of various transplantable tumor cell lines. The degree of inhibition is dependent upon the form of selenium administered and the quantity given. This presentation will discuss some of the aspects of route of administration, dose, and form of selenium on the degree of tumor inhibition observed with various cell lines. Much of the work in our laboratory has been conducted with the Ehrlich ascites tumor cell and the L1210 leukemic cell line. Part of the discussion will also emphasize the importance of antagonists as factors that may modify the beneficial effects of selenium supplementation. Some of the possible mechanisms by which selenium may retard transplantable tumor cells will also be discussed.

More details were provided by Dr. Milner and his colleague, Dr. Glenn Greeder, in the August 15, 1980 issue of *Science* (pp. 825–27). They concluded, "Selenium, administered to mice with Ehrlich ascites tumors effectively limited tumor growth. The response was dependent on the chemical form and dose of selenium administered. At the doses administered, there were no detectable adverse effects to the host."

At the British Columbia Cancer Research Centre and Department of Medical Genetics in Vancouver, Canada, Drs. R. Whiting, L. Wei, and H. Stich reported on the antimutagenic activity of natural organic selenium compounds:

> Selenocystine has potent *antimutagenic* activity. When tested against model mutagenic carcinogens, the protective effect of selenocystine exceeded that of sulfur (cysteine) and oxygen (serine) analogs. Selenomethionine did not have appreciable antimutagenic effects. Thus specific organic selenium compounds may act physiologically to protect against mutagens and carcinogens without being themselves mutagenic.

Selenium is also involved in the healthy continuance of the body's second line of defense—the liver. The manner in which selenium protects the liver was originally studied by Drs. Klaus Schwarz and Milton Scott in their attempts to determine the nutritional need for selenium. Since the liver detoxifies carcinogens, it follows that a selenium deficiency will impair the liver's ability to destroy carcinogens.

Selenium stimulates the immune response—the body's major line of defense. Some scientists believe that under normal circumstances people develop precancerous cells that are destroyed by the immune system before they can bloom into full-fledged cancers. Those who suffer from a weakened immune system naturally lose this protection. This role of selenium will be discussed further in Chapter Six.

Treating Cancer

In 1956, four leukemia patients were given an organic selenium compound called selenocystine. In all four instances, a rapid decrease of the total leukocyte count was observed, as well as a decrease in spleen size. The most striking results, however, were observed in the cases of acute leukemia.[27]

In 1975 Dr. K.P. McConnell noticed a link between the survival times of 110 cancer patients and their blood selenium levels. Dr. McConnell and his colleagues concluded that those patients with the lowest blood selenium levels were more likely to have far-spreading cancer, multiple tumors located in different organ systems, and multiple recurrences.[28,29] Conversely, those patients whose cancers were confined and who rarely suffered recurrences all had higher (but still subnormal) selenium levels. These same researchers determined in 1978 that the blood selenium levels of breast cancer patients were lower than women in the same age brackets without breast cancer.[30]

At the Second International Conference on Inorganic and Nutritional Aspects of Cancer, held in La Jolla, California on January 5, 1979, Dr. Gerhard Schrauzer commented: "Selenium supplementation promises to become a method of cancer prophylaxis applicable at the individual or community level. Selenium supplementation is recommended particularly for individuals at high risk. This includes those with familial predisposition to cancer. Workers in the chemical and metal industries could also benefit from selenium supplementation."

In addition, Dr. Schrauzer discussed the potential therapeutic value of selenium and its use as an adjuvant in cancer chemotherapy. Dr. Schrauzer has monitored the blood levels of selenium in cancer patients and has observed that cancer patients seem to have lower blood levels of selenium produced by a given dietary portion.

Earlier, at the Symposium on Selenium-Tellurium in the Environment, held at the University of Notre Dame on May 13, 1976, Dr. Schrauzer reported that in cancer patients blood selenium levels are directly related to survival time and inversely related to recurrences and metastases.

At this writing, a possible breakthrough may be developing. At least three physicians have observed that cancer patients respond better to treatment when their blood level of selenium is raised to a critical level. I am not at liberty to release their names nor their hospital until after their results are published in the medical literature. The first paper should appear in late 1980.

One physician gave selenium as a nutritional adjunct to conventional therapy to over 100 patients and reported that those patients generally did much better than expected. As an example, of eight patients with inoperable lung cancer, four were alive and well one year later, with no progression of the disease as shown by X-ray examination. Normally, all eight would have been expected to have died in that time.

The doctor gave the patients up to 2000 micrograms of selenium daily, along with other antioxidant nutrients including vitamins A, C and E. He would adjust the selenium supplementation to produce normal blood selenium levels which often did not occur until 900 to 2000 micrograms per day were given. He found no evidence of nerve, liver or blood abnormalities (due to possible selenium toxicity) in any patient, nor in autopsies of thirty-seven patients.

Many questions remain concerning selenium and its positive effects on cancer that need answering. But there is no scientific doubt now that cancer is linked to a dietary selenium deficiency. Be sure you are not selenium deficient!

SUMMARY

Selenium is protective against cancer and selenium-deficient persons may have reduced chances of successful cancer therapy.

- [ ] Selenium may reduce breast cancer to 80 percent of its present level.
- [ ] The ideal selenium intake to protect against cancer may be 300 micrograms daily.
- [ ] More cancers occur in low-selenium areas.
- [ ] Residents of cities with low selenium blood levels have higher cancer death rates.
- [ ] Selenium protects animals that have been fed cancer-causing compounds from getting cancer.
- [ ] Selenium added to the diets of cancer patients has improved their survival time after therapy.

REFERENCES

1   Scarr, L. *The San Diego Union*, January 10, 1979.

2   Workshop on chemoprevention of cancer, National Cancer Institute. Bethesda, Maryland. February 2, 1978.

3   Stiller, R. *Family Circle*, June 14, 1978.

4   Mishara, E. *National Enquirer*, April 4, 1978.

5   Schrauzer, G. 1978. *Inorganic and Nutritional Aspects of Cancer*. New York: Plenum Press, p. 336.

6   Schrauzer, G. and Rhead, W.I. 1971. *Experientia* 27:1069–71.

7   Schrauzer, G., Rhead, W. and Evans, G. 1973. *Bioinorganic Chemistry* 2:329–40.

8   Shamberger, R. and Frost, D. 1969. *Canadian Medical Association Journal* 100:682.

9   Shamberger, R. and Willis, C. 1970. *Journal of the National Cancer Institute* 44:931.

10 Shamberger, R. and Willis, C. June 1971. *CRC Critical Reviews in Clinical Laboratory Sciences*, pp. 211–21.

11 *Chemical and Engineering News*, January 17, 1977, p. 35.

12 Schrauzer, G. 1978 *Inorganic and Nutritional Aspects of Cancer*. New York: Plenum Press, p. 334.

13 Schrauzer, G., White, D. and Schneider, C. 1977. *Bioinorganic Chemistry* 7:36.

14 Shamberger, R. and Willis, C. 1971. *Critical Review of Clinical Laboratory Science* 2:211–21.

15 Wedderburn, J. 1972. *New Zealand Veterinarian Journal* 20:56.

16 Clayton, C. and Bauman, C. 1949. *Cancer Research* 9:575–82.

17 Pryor, W. 1966. *Free Radicals*. New York: McGraw-Hill.

18 Harr, J., Exon, J., Whanger, P. and Weswig, P. 1972. *Clinical Toxicology* 5(2):187–94.

19   Passwater, R. 1973. *American Laboratory* 5(6):10–22. (Also U.S. pat. appl. 39140, 97011, 271655, 398596, 481788)

20   Wattenberg, L. 1972. *Journal of the National Cancer Institute* 48:1425–31.

21   Jacobs, M. and Griffin, C. 1977. *Cancer Letters* 2:133–38.

22   Griffin, C. and Jacobs, M. 1977. *Cancer Letters* 3:177–81.

23   Schrauzer, G. and Ishmael, D. 1974. *Annals of Clinical Laboratory Science* 4:441–7.

24   Whanger, P., Tinsley, I., Schmitz, J. and Exon, J. May 1980. Second International Symposium on Selenium in Biology and Medicine, Texas Tech University, Lubbock, Texas.

25   Schrauzer, G. 1978. *Inorganic and Nutritional Aspects of Cancer.* New York: Plenum Press, p. 330.

26   Jansson, B. 1975. *Cancer* 36:2373–84.

27   Weisberger, A. and Suhrland, L. 1956. *Blood* 11:19.

28   McConnell, K., Broghamer, W., Blotcky, A. and Hurt, O. 1975. *Journal of Nutrition* 105:1026–31.

29   Broghamer, W., McConnell, K. and Blotcky, A. 1976. *Cancer* 37:1384.

30   McConnell, K. 1979. *Advances in Nutrition Research,* vol. 2, ed. by H. Draper. New York: Plenum Press, p. 225.

THREE

Selenium builds a healthy heart

SELENIUM'S role as a protector against cancer is dramatic, but its protective role against heart disease—our number one killer—is selenium's greatest value. The evidence supporting selenium's role in relationship to heart disease seems even more convincing than it does for cancer.[1] Whether or not a selenium deficiency causes heart disease is unproven. What seems certain is that a selenium deficiency increases the susceptibility to heart disease.

The leading cause of heart disease death is myocardial infarction, which is the death of heart tissue due to the lack of blood in a region of the heart. This lack of blood is usually caused by a blood clot (thrombus) in a coronary artery; thus the usual precipitating event that injures the heart is called coronary thrombosis.

Much attention has been given to atherosclerosis, the development of arterial plaques (normally thought of as cholesterol deposits), but several factors are equally important to a healthy heart. The blood must maintain its normal slipperiness and resist the tendency to clot. The arteries must remain open to allow adequate blood volume through—rather than be obstructed with plaque. And—what is often overlooked completely—the heart itself must maintain muscle fiber strength.

It is interesting to note that two of the three factors involve muscle fibers. The artery is principally muscle fiber; and the media (central artery layer) is smooth muscle fiber. The muscle is required to hold artery shape and to withstand the pressure of the heartbeat. The plaque that contains cholesterol actually begins as a mutated muscle cell in the media and grows into the artery's intima (innermost) layer as it proliferates.[2,3,4,5,6,7] This is called monoclonal proliferation and will be discussed later in this chapter.

The calcification (calcium deposition) of the arterial plaques produces hardening of the arteries (arteriosclerosis). Calcium deposits can also form in the heart, which then takes on a white, chalk-like appearance. Animal nutritionists have noted that animals living in selenium-deficient areas developed such calcification of their hearts; the disease was named white-muscle disease.[8]

My early involvement with selenium and heart disease centered on selenium's role in muscle health. Animal nutritionists had determined that selenium deficiencies caused nutritional muscular dystrophy (see Chapter Nine) and a degeneration of skeletal muscle called Zenker's disease. A similar degeneration of the Purkinje fibers which cause the heartbeat had been observed in selenium deficiency. In fact, the hearts of most selenium deficient animals will collapse when surgically removed, while hearts from adequately nourished animals will hold their shape.

The role of selenium as an antioxidant suggested to me that this vital element protected the muscle cells against free-radical attack. I also reasoned that selenium would reduce the free-radical attack on the arterial wall, thus reducing arterial plaque formation.

Dr. Earl Benditt proved that the plaques were due to monoclonal cell proliferation, which could be caused by free radicals. My research was directed towards the theory that the free radicals caused injury to the cells which initiated a repair process that pulled cholesterol out of the bloodstream. Of course, I had trouble establishing that exact mechanism. I was close, but not close enough. Fortunately, Dr. Benditt did establish the correct mechanism of free-radical injury to a smooth muscle cell which then multiplies rapidly to form the plaque. The plaque of smooth muscle cells then manufacture cholesterol, other lipids (fats) and collagen.

Even though my hypothesis was not verified, the protective effect of selenium was demonstrated and I described this in my 1970 patent applications. Since that time, other scientists have found additional, perhaps even more important, ways in which selenium protects against heart disease.

Before we consider the human data, let's take a look at the early animal studies that paved the way for the human studies. We owe so much to this pioneering work.

In 1965, Dr. K. O. Godwin found that rats fed a low selenium diet developed abnormal electrocardiograms.[9]

Later he determined that lambs fed low selenium diets also developed abnormal electrocardiograms as well as blood pressure disturbances.[10] He confirmed these findings in animals grazing on low selenium pastures.[11] In addition to electrocardiogram and blood pressure abnormalities, these animals also had cardiac muscle lesions and circulatory disorders.

Dr. Milton Scott was one of the members of Dr. Klaus Schwarz's research team that studied selenium in animal health in the late 1950s. He later moved to Cornell University, where he uncovered a wealth of knowledge about the roles of selenium, vitamin E and the sulfur-containing amino acids.

Dr. Scott found that selenium-deficient chickens and turkeys quickly developed heart degeneration, even before developing skeletal muscle or liver degeneration.[12] The selenium-deficient hearts also typically developed Zenker's degeneration and hemorrhages (see Figure 3.1).

Ranchers realized the economic loss they were suffering as a result of losing young livestock from white muscle disease. In 1968, Dr. David Carter, research soil scientist for the U.S. Department of Agriculture at the Idaho Agriculture Experiment Station prepared an information sheet to inform ranchers that white muscle disease is caused by a selenium deficiency.[13] Dr. Carter advised the owners to provide animal feed with protective selenium concentrations or to inject young animals with selenium soon after birth.

More than half of the poultry in the United States is produced in selenium deficient states. It is estimated that poultry producers suffer annual losses of more than 27 million dollars due to selenium deficiency.[14] It is also estimated that losses to the swine industry equal the losses in the poultry industry, so the annual losses due to sele-

FIG. 3.1
Heart degeneration due to selenium deficiency

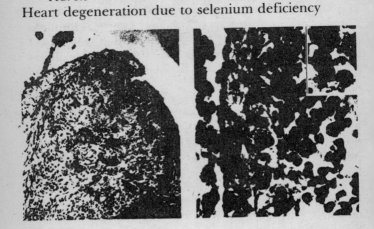

Left: Heart, left atrium. Acute Zenker's degeneration with muscle bundles separated by edema and hemorrhages. (Basal diet, three weeks.)

Right: Heart, left atrium. Erythrocytes in left portion of field are outside epicardium. Extensive subepicardial hemorrhage and edema with admixture of heterophilic leukocytes.
Inserted in upper right corner: detail of heterophilic leukocytes. (Basal diet, three weeks.)

Source: Scott, M., Olson, G., Krook, L. and Brown, W. 1967. *Journal of Nutrition* 91 (4):581.

nium deficiency in swine and poultry in the U.S. are about 55 million dollars. Additional losses to sheep, cattle, race horses, mink and goats (due to selenium deficiency) would most likely push the total U.S. agricultural loss far beyond 100 million dollars. In 1973, Canada approved the addition of selenium to poultry and swine feeds, and U.S. approval followed in February 1974.[15]

What is the dollar loss to human life? It is unthinkable to discuss human life in terms of dollars, but I believe the hospitalization costs due to avoidable illness caused by selenium deficiency would easily exceed the agricultural loss.

In swine, the heart destruction brought about by selenium deficiency is called "mulberry heart disease." Pigs eating a selenium-deficient diet usually die between two and four months of age. Their hearts appear reddish-purple (similar to a mulberry) because of extensive bleeding from the smaller blood vessels of the heart.[16,17, 18]

This is similar to the disease afflicting young children in certain areas of China, where, as mentioned in Chapter One, it is called "Keshan disease," which will be discussed below.

In 1971, Dr. L. Sprinker and his associates at Oregon State University concluded that "the primary nutritional role of selenium may be in the growth and maintenance of the vascular beds and membranes."[19] They have confirmed that selenium deficiency in the rat leads to heart muscle damage.

A Soviet research team discovered that selenium was needed to minimize the damage caused by heart attacks.[20] They blocked the flow of blood in a coronary artery and measured the extent of damage. Extra selenium supplementation decreased the extent of heart damage and prevented the appearance of gross changes on the electrocardiogram.

Researchers at the University of Tennessee found that when both vitamin E and selenium were given to rats for 180 days prior to causing a heart attack by tying off a coronary artery, the damage (myocardial necrosis) was slight compared to that in animals receiving only one or none of the two nutrients.[21]

Human studies

Oddly enough, the very fact that selenium plays such a vital role in preventing cancer, heart disease and other diseases has hampered the study of selenium itself. People tend to label or tag an item to conveniently categorize it in the general scheme of things. However, items are not necessarily limited to our understanding of them. For example, a screwdriver can do more than turn a screw—it can pry open a lid, scrape paint and even open envelopes.

By the same token, ascorbic acid is more than vitamin C. When we think of it as a vitamin, we tend to limit our expectations of ascorbic acid to the known function of classical vitamins. However, when we recognize ascorbic acid or tocopherol (vitamin E) as compounds having wide utility in addition to vitamin activity, then we remove our mental "blinders."

Dr. Shamberger had evidently initially categorized selenium as a nutrient also involved in cancer prevention. In the late 1960s, after establishing the relationship of selenium deficiency to greater risk of cancer, he also determined a similar relationship with heart disease. The animal research that we just reviewed was not widely disseminated at that time. Lacking a clear animal model linking selenium deficiency to heart disease, Dr. Shamberger filed away his data while his research group pursued their cancer research. Dr. Helen Cannon also noticed this link and reported it to Dr. Douglas Frost in 1968.[22]

As more and more researchers published their studies linking selenium deficiencies in various animals to heart disease, Dr. Shamberger retrieved his filed data and presented his evaluation of it in 1976 at the annual meeting of the Federation of American Societies for Experimental Biology in Anaheim, California. There, along with Dr. Charles Willis, Dr. Shamberger reported that Americans living in selenium-deficient areas are three times more likely to die from heart disease as those living in selenium-rich areas. Shamberger and Willis showed that the heart disease rate in the fifty-five- to sixty-four-year-old age group was lowest in the selenium-rich states; Texas, Oklahoma, Arizona, Colorado, Louisiana, Utah, Alabama, Nebraska, and Kansas. Colorado Springs, Colorado, was 67 percent below the national average for heart disease deaths and Austin, Texas, was 53 percent below.

In the selenium-deficient states—notably Connecticut, Illinois, Ohio, Oregon, Massachusetts, Rhode Island, New York, Pennsylvania, Indiana, Delaware and the District of Columbia—the heart disease death rate was substantially higher than the national average. In the District of Columbia, for example, the heart disease death rate was 22 percent above the national average.

Table 3.1 represents a clear picture of the relationship between selenium availability and heart disease.

Of course, epidemiological studies only show association and are not proof of cause and effect. That is provided by considerable other research.

Dr. Shamberger suggests, "If selenium has an effect on heart disease—and the evidence points to it—then Americans are probably 100 to 150 micrograms short in their selenium intakes. A supplementation of the American diet could bring about the desired increase of the total intakes to about 250 to 350 micrograms per day."[23]

TABLE 3.1
World heart disease rates versus selenium intake

Country	Selenium Intake micrograms/day	Coronary Heart Disease Rate
Finland	25	1009
USA	61	870
Canada	62	722
Ireland	75	722
Australia	76	867
Norway	82	602
Greece	92	236
Poland	94	301
Yugoslavia	99	232
Bulgaria	108	331

Selenium intake and coronary heart disease deaths per 100,000 in 55- to 64-year-old males in ten countries.

Source: Data with exception of Finland from Shamberger, R., Gunsch, M., Willis, C. and McCormack, L. 1975 *Trace Substances in Environmental Health* ix ed. D. Hemphill. Columbia: University of Mississippi Press, pp. 15–22. Author has corrected the Finnish figures to that used by Dr. Johan Bjorksten and Dr. Pekko Koivistoinen, head, Dept. of Food Chemistry and Technology, University of Helsinki.

Dr. Shamberger elaborates on his research findings and provides helpful advice in Chapter Thirteen.

Dr. Johan Bjorksten, president of the Bjorksten Research Foundation of Madison, Wisconsin, has found that heart attacks are seven times more prevalent in low-selenium areas of Finland than in high-selenium areas.[24,25] He also noticed that the highest cardiac infarction death

TABLE 3.2

U.S.A. heart disease death rates and selenium

Disease	Selenium levels			
	Very high	High	Medium	Low
MALES				
Coronary	774	818	893	962
Hypertensive	34	53	64	71
Cardiovascular Renal	1045	1149	1252	1308
Cerebrovascular	108	138	159	139
FEMALES				
Coronary	220	225	249	306
Hypertensive	27	39	47	53
Cardiovascular Renal	413	428	474	539
Cerebrovascular	89	94	109	104

Age specific death rates per 100,000 for white males or females, age 55–64, 1959–1961. Selenium level classification, above 0.26 is very high, 0.10 to 0.25 is high, 0.06 to 0.09 is medium, and 0.01 to 0.05 is low.

Source: Shamberger, R. May 11–13, 1976. *Proceedings of the Symposium on Selenium-Tellurium in the Environment.* Univ. Notre Dame, p. 265.

rates occurred in the two countries—Finland and New Zealand—with the lowest amounts of soil selenium. He then compared the heart attack death rates of working age persons (fifteen to sixty-four years) in various Finnish counties. In those counties with 0.1 ppm selenium in the drinking water, he found one heart attack for every 1,730 residents of working age. In those counties having less than 0.05 ppm selenium, Dr. Bjorksten uncovered one heart attack for every 224 residents of working age.

Selenium supplements and their effects on some form of heart disease in China

China has a huge low-selenium belt running from the northeast to the southwest. A form of heart disease called "Keshan disease," is prevalent in this low-selenium area. Keshan disease resembles the Mulberry heart disease in selenium-deficient swine. The people most susceptible to this disease are children and women of childbearing age.

A study of the effect of selenium supplements on Keshan disease was initiated in 1974. In Mianning county, selenium supplements were given to 4,510 children selected at random, while 3,985 others made up the control group receiving the placebo. The following year these two groups were increased to 6,709 and 5,445, respectively. The results were so dramatic that the control group was abolished in 1976 and all children (who now numbered 12,579) were given selenium supplements. Thus, 99 percent of the children aged one to nine in four communes in the county participated in the clinical trial.

As reported in the *Chinese Medical Journal* and *Lancet*: "In 1974, of the 3,985 children in the control group, there were fifty-four cases of Keshan disease (1.35 percent), while only ten of the 4,510 selenium supplemented children fell ill (0.22 percent). The difference in morbidity rate between the two groups was highly significant.

"Again a significant difference was shown in the 1975 figures with fifty-two of 5,445 children in the control group (0.95 percent) and only seven of the 6,767 in the treated group (0.1 percent).

"As the result of these two years showed that oral administration of selenium had positive effects in the prevention of Keshan disease, all the children were given selenium supplements from 1976 on. In consequence, only four cases occurred out of the 12,579 children in 1976, further lowering the rate to .03 percent. *In 1977, there were no fresh cases among the 12,747 treated children.*"[26]

The Chinese are seriously considering adding selenium to table salt as a protective measure. We added iodine to salt to prevent goiter but as yet we have not decided to add selenium as a routine preventive.

Selenium is vital to the heart

Coenzyme Q (also called ubiquinone) is an aid in certain energy-producing enzymatic reactions and is indispensable to heart function.[27] Since the normal human heart is higher in Coenzyme Q content than most other tissues,[28] it follows that biopsy samples taken from the hearts of heart disease patients show a deficiency of Coenzyme Q.[29]

Animal studies have shown that a deficiency in Coenzyme Q lessens heart vitality and produces cardiac degenerative lesions. Selenium appears to be the most important nutrient in the control of Coenzyme Q levels. Therefore, adequate selenium is required to produce the necessary Coenzyme Q required for a healthy heart.

Selenium controls blood pressure and plaque formation

Selenium normalizes blood pressure by controlling prostaglandin production. Prostaglandins are a family of chem-

ical messengers, much like hormones, that control many
bodily functions.

Dr. J. E. Vincent of the Department of Pharmacology
at the Erasmus University, Rotterdam (The Netherlands)
believes that a dietary selenium deficiency results in the
inability of the body to produce certain prostaglandins.
The resulting deficiency of prostaglandin E_2 makes the
blood stickier and thus more likely to clot. The production of prostaglandin A_2 is also diminished, which results
in high blood pressure. Dr. Vincent has indicated that a
selenium-containing enzyme is required to manufacture
several prostaglandins. If a selenium deficiency occurs,
not only does a prostaglandin deficiency result, but several compounds that were intended to be incorporated
into prostaglandins accumulate in the blood to encourage
degeneration of the arteries and induce platelet aggrega-
tion.[30]

Animal studies conducted by Dr. James Aiken of the
Upjohn Laboratories indicate that a prostaglandin released
by the kidney may be the body's most important regula-
tor of blood pressure.

Selenium levels in the kidneys of patients with high
blood pressure are usually lower than in normal controls.[31] Furthermore, Dr. H. Perry of the University of
Missouri has found that selenium partly detoxifies (neu-
tralizes) the adverse effect of the mineral cadmium, which
induces high pressure in laboratory animals.[32] Dr. Shamberger has also noted that geographical areas with high
amounts of cadmium produce above normal heart disease death rates.[33]

Selenium is part of the enzyme glutathione peroxidase, which inhibits aberrant oxidations that damage cellular membranes. Glutathione peroxidase protects cellular
membranes against attack by free radicals that could cause

the cell to proliferate wildly, similar to a cancer, and cause arterial plaque (deposits) formation.

Dr. L. H. Sprinker of the Department of Veterinary Medicine of Oregon State University and his colleagues reported that selenium deficiency in rats produced plaques and abnormal vascularization associated with endothelial degeneration in their arteries.[34]

Selenium used successfully as a drug

The drug Tolsem (Chromalloy), which consists of vitamin E and selenium, has been given a two-year clinical trial in Mexico. This combination was said to have demonstrated 92 percent beneficial responses in patients with recurring attacks of angina pectoris—with or without myocardial infarct.[35,36] These responses involved reduction or elimination of angina attacks, increased vigor and work capacity, and improved electrocardiograms. There was no evidence of adverse side effects.

SUMMARY

Epidemiological evidence, animal research, and theoretical biochemistry all support the premise that selenium deficiency increases the susceptibility to heart disease.

☐ The heart muscle requires selenium.

☐ Animals deficient in selenium develop heart disease similar to human heart diseases.

☐ Selenium supplements reduce damage after induced heart attacks.

☐ Heart disease death rates are higher in low-selenium areas.

☐ Heart disease in one area of China was virtually eliminated with selenium supplements.

☐ Selenium helps regulate blood pressure.

☐ Selenium protects against arterial deposits.

☐ Selenium supplement reduces heart pain.

REFERENCES

1 Frost, D.V. *Bulletin of Selenium-Tellurim Development Association* no. 17.

2 Passwater, R. 1977. *Supernutrition for Healthy Hearts*. New York: Dial Press, pp. 5–6, 27–36, 312–16.

3 Moss, N.S. and Benditt, E.P. 1970. *Laboratory Investigation* 23:231–45 and 521–35.

4 Poole, J.C.F., Cromwell, S.B. and Benditt, E.P. 1970. *American Journal of Pathology* 62:391–404.

5 Schwartz, S.M. and Benditt, E.P. 1973. *Laboratory Investigation* 28:699–701.

6 Benditt, E.P. and Benditt, M.M. 1973. *Proceedings of the National Academy of Science* 70:1753–6.

7 Vracko, R. and Benditt, E.P. 1974. *American Journal of Pathology* 75:204–7.

8 Hartley, W. and Grant, A. 1961. *Federation Proceedings* 20:679.

9 Godwin, K. 1965. *Quarterly Journal of Experimental Physiology* 50:282.

10 Godwin, K. and Fraser, J. 1966. *Quarterly Journal of Experimental Physiology* 51:94.

11 Godwin, K. 1968. *Nature* 217:1275.

12 Scott, M., Olson, G., Krook, L. and Brown, W. 1967. *Journal of Nutrition* 91(4):581.

13 Carter, D. June 1968. *Current Information Series* no. 84. College of Agriculture, Univ. Idaho.

14 Van Houweling, C. April 19, 1973. Environmental Impact
    Statement. *Rule Making on Selenium in Animal Feeds.* Bureau
    of Veterinary Medicine F.D.A., Dept of HEW.

15 1974. *Federal Register* 39 (5):1371.

16 Grant, C. 1961. *Acta Veterinary Scandinavia* 2 (Supplement
    3).

17 Cunha, T. 1969. *Feedstuffs* 41(18):21.

18 Ruth, G. and Van Vleet, J. 1974. *American Journal of Veteri-
    nary Research* 35(2):237–44.

19 Sprinker, L., Harr, J., Newberne, P., Whanger, P. and
    Weswig, P. 1971. *Nutrition Reports International* 4(6):335–39.

20 Nikolaev, S., Kudrin, A. and Kaktursky, L. 1976.
    Farmakologia Toksikologia 39(5):571–74.

21 Revis, N. and Armstead, B. 1979. *International Congress
    Series—Excerpta Medica.* 491:1029–32.

22 Frost, D. March 1980. *CRC Critical Review of Toxicology*
    15(6).

23 Private communication, Executive Health Report, 1980,
    and May 3, 1976. *Chemical and Engineering News*, p. 25.

24 Bjorksten, J. 1979. *Rejuvenation* 7(3):61–66.

25 September 24, 1979. *Chemical and Engineering News*, p. 24.

26 1979. *Chinese Medical Journal* 92(7):471–76. Also see Octo-
    ber 27, 1979. *The Lancet*, p. 890.

27 Folkers, K. 1970. *Feedstuffs.* Also see 1965 Kummerow, F.,
    ed. *Metabolism of Lipids.* Springfield, Illinois: Charles C.
    Thomas, P. 262.

28 Chipperfield, B. 1971. *Clinical Chim Acta* 31:459–65.

29 Edwin, E., Diplock, A., Bunyan, J. and Green, J. 1971.
    Biochemistry Journal 125:407.

30 Vincent, J. 1970. *Prostaglandins* 8(4):339–40.

31 Larsen, N. et al. 1972. *Nuclear Activation Techniques in the
    Life Sciences.* Vienna, Austria: International Atomic Energy
    Agency, p. 568.

32 Perry, H. et al. 1974. In *Trace Substances in Environmental
    Health. Proceedings, Univ. of Missouri* 8:51. ed. D. Hemphill.

33 *Chemical and Engineering News*, May 3, 1976, p. 25.
34 1971. *Nutritional Reports International* 4(6):335–39.
35 *Wall Street Journal*, March 13, 1973.
36 Willalon, J. 1974. Thesis, Univ. Michoacana de San Nicolas de Hidalgo, Morelia, Mexico.

FOUR

Slowing down the aging process with selenium

Nᴏᴛ only is selenium protective against cancer and heart disease, it slows the aging process because the fundamental mechanism of all three is the same. In fact, it was while I was studying the role of selenium in retarding the aging process that I noted the reduced incidence of cancer in my experimental animals.

This is not to imply that taking extra selenium will make you younger. It does mean that if you fail to get your optimum amount of selenium, you may become "older than your years." Your body will not have full protection against aging factors and you will look older and physiologically be older than you would be otherwise.

Retardation of the aging process and protection from disease are interrelated. In longevity experiments, it's difficult to determine if an increased lifespan represents a slowing of the aging process or the avoidance of life-shortening diseases. There is a strong correlation between most types of cancer and age. It could be that cancer appears in older people because it takes decades for some cancers to develop. It could also be caused by an increased exposure to carcinogens as the years roll by. However, recent studies indicate that the aging process might make people more vulnerable to cancer.[1-4] Thus if we slow the aging process, we are less vulnerable to cancer and most other diseases.

Indeed, selenium does help us to live better longer. My first interest in selenium involved my experiments on retarding the aging process in the late 1960s and early 1970s. I chose selenium as a compound that might be synergistic with other compounds previously shown to extend the lifespan.

Dr. Denham Harman had previously shown that vitamin E and a few synthetic antioxidants extended the average lifespan of mice.[5-7] The antioxidants used in Dr. Harman's experiments were essentially fat-soluble compounds used in amounts approaching one-half percent of the diet. I reasoned that a combination of antioxidants that were active in all body systems—not just fatty tissue systems—would be more effective. I also hypothesized that compounds having interrelated multiple functions would provide further life extension at more reasonable amounts.

I chose selenium because various selenium compounds were water-soluble antioxidants. Dr. Milton Scott of Cornell University had shown that selenium was interchangeable with some of the functions of vitamin E.[8] Dr. Al Tappel of the University of California at Davis, demonstrated in 1964 that several naturally occurring selenium compounds could protect biological systems against radiation.[9] In 1965, Dr. Tappel showed that selenocystine (a selenium containing amino acid) protected certain key enzymes against unwanted oxidation.[10] I used selenocysteine and dimethyl selenide as my sources of selenium. The use of selenium as an antioxidant, radiation protector and partner of vitamin E justified its inclusion in the combination of antioxidants tested with a series of *in-vitro* (test-tube type) experiments in 1965 to measure the anti-radical activity (ability to quench deleterious free radicals). The *in-vitro* experiments, ingeniously designed by Dr. Johan Bjorksten, used the polymerization of gelatin as a model for the aging of body tissues.[11]

The amount of selenium required to produce synergism in the *in-vitro* model suggested that the safety range before reaching toxicity was small. Several researchers had found that amino acids containing sulfur were protective against selenium toxicity.[12,13] Since sulfur-containing amino acids are also effective antioxidants, I included them in my longevity experiments with mice. By 1969, I had successfully developed a range of combinations of antioxidant compounds that extended the mean lifespan by 20 to 30 percent and the maximum lifespan by 5 to 10 percent.[14-16] The December 1971 issue of *Prevention* magazine carried a review of my work.[17]

By 1971, I found that these combinations produced lifespan increases of 175 percent compared to controls, in which the mice were given aging accelerants.[18-20] When cancer-causing compounds were given to the animals, it was found that the antioxidant combinations reduced the cancer incidence from one-sixth to one-tenth of that observed in the control animals. As a result of these and other experiments, we now understand a great deal more about the role selenium can play in keeping us younger longer.

What is aging?

The American Medical Association's Committee on Aging has studied the problem of human aging for more than a decade. And in that time, the committee has not found one physical or mental condition that can be directly attributed simply to the passage of time. Some of the alleged diseases of aging—such as high blood pressure and arthritis—are prevalent in the very young as well as the very old. What exactly is aging then, and what are its causes?

Aging can be described as the process that reduces the number of healthy cells in the body. Although we have noted the increase of some enzymes in the body, and the decrease of others, the most striking factor in the aging process is the body's loss of reserve due to the decreasing amount of cells in each organ. For example, fasting blood glucose levels remain fairly constant throughout life, but the glucose tolerance measurement, which measures the reserve capacity of this system to respond to the stress of the glucose load, shows a loss of response in aging. The same holds true concerning the recovery mechanisms of other parameters.

Cellular aging actually begins before birth and is the one factor underlying the aging process of the entire body. The stability of the living system becomes progressively impaired by chemical reactions, not the passage of time. If we can control the rate of these deleterious reactions, therefore, we can control the advance of physiological aging.

Free-radical reactions produce five different types of reactions that result in the body's loss of active cells. Since the cumulative effect of billions of cellular free-radical reactions can add to the body's loss of reserve, it follows that free radicals can accelerate the aging process.

After studying the aging process for the past twenty years, it is my opinion that what we have come to recognize as the signs of aging are the result of two basic processes. One is the fundamental aging process itself, and the second is a process determined by the environment—i.e., each person's status and life-style. The combined effects of both processes determine our *apparent* age—i.e., the age others assume us to be.

Environmental factors typically cause aging signs to appear ten to twenty years earlier than they would under normal circumstances. There is evidence to suggest that many signs of old age are due solely to environmental factors and are not even related to aging. Older people simply have had more opportunities to abuse their bodies through suboptimal nutrition and "lazy" life-styles.

Genetics certainly plays the major role in determining how we age. What people really notice about aging, however, is the difference between the individual and the average. You can appear to be years younger just by minimizing the environmental aspects of aging. By nourishing your body optimally and by avoiding stress beyond the body's reserve capabilities, you can slow down the rate of aging. The combined effects of the primary and secondary aging processes will therefore be essentially the same as the primary rate alone, which will give you a tremendous advantage over the vast majority that accelerate their aging and hasten their demise. Of course, they won't see it that way—they will think that they are normal and you are slowing your aging process.

Free radicals

Often I have discussed the role of free radicals in cancer, heart disease and aging. Let's take a more detailed look at free radicals to better understand how they cause such extensive, and seemingly unrelated damage.

As we have seen, a free radical is an incomplete molecule. It is a fragment of a molecule that is highly reactive, because its electron arrangement is out of balance. Atoms, molecules and ions are more stable entities because they have more balanced electron arrangements.

You may want to read Appendix A which explains free radicals.

The highly reactive free radicals do more damage than that of one molecule to one molecule reaction. Each free radical is capable of destroying an enzyme or protein molecule or destroying an entire cell. However, the damage is much more extensive than that because each free radical usually generates a chain of free radical reactions resulting in thousands of free radicals being released to destroy body components.

Dr. William Pryor of Louisiana State University points out ways in which free radicals do extensive damage to our bodies. "This biological magnification occurs for two reasons. The first, and most important, is the enormous sensitivity of the cell to modifications in its heredity apparatus such as its DNA. The chromosomes, which control the reproduction of the cell, are extremely radiation sensitive; the cytoplasm is much less so. Largely because of the sensitivity of DNA, radiation that destroys only one molecule in one million or ten million in the cell can be lethal.

"The second cause of biological magnification is that any polymeric system is sensitive to small chemical changes, and many important biomolecules are polymers."[21]

Free-radical reactions leading to cell membrane damage can cause cancer, heart disease or accelerated aging. Drs. Denham Harman and Al Tappel have extensively described the role of free radicals in aging.[22-27] Figures 4.1 and 4.2 illustrate a few of the deleterious free-radical reactions that accelerate the aging process.

There are five basic types of damage caused by free radicals that accelerate aging. (A more detailed description of the free-radical theory of aging appears in the appendix).

1. Lipid peroxidation, in which free radicals initiate damage to fat compounds in the body, causing them to turn rancid and release more free radicals.

2. Cross-linking, in which free-radical reactions cause proteins and/or DNA to fuse together (see Figure 4.1).

3. Membrane damage, in which free-radical reactions destroy the integrity of the cell membrane, which in turn interferes with the cell's ability to bring in nutrients and expel wastes.

4. Lysosome damage, in which free-radical reactions rupture lysosome membranes; these then spill into the cell and digest critical cell compounds (see Appendix A and Figure 4.2).

5. Accumulation of the age pigment (lipofuscin), which may interfere with cell chemistry.

The most damaging agents of free-radical reactions include the superoxide radical (O_2^-), hydroxyl radical ($OH\cdot$), lipid peroxide radical ($LOO\cdot$) and hydrogen peroxide (H_2O_2).

FIG. 4.1

Abnormal cross-linkage of molecules caused by free radicals

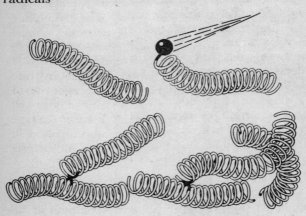

Abnormal cross-linkage of molecules caused by free radicals

Molecular havoc occurs when the normal orderly arrangement of enzymes and other components is disturbed.

Source: Reproduced with permission of *Nutrition Today* magazine. P.O. Box 1829, Annapolis, Maryland 21404. © December 1967.

The body defends itself against these agents with superoxide dismutase (an enzyme that destroys the superoxide radical), catalase (an enzyme that produces vitamin E (a general antiradical) and glutathione peroxidase (an enzyme that stops lipid peroxidation and converts hydrogen peroxide to water). Each molecule of glutathione peroxidase contains four atoms of selenium (see Figure

4.3). Thus selenium is a key component of the body's defense against accelerated aging.

Dr. L. Flohé of the Chemie Grunenthal GmbH in the Federal Republic of Germany has shown that glutathione peroxidase protects cells from mutagenic peroxides formed from DNA and nucleotides. It also breaks down lipid (fat) peroxides that can contribute to atherosclerosis. [28] It is also involved in the regulation of carbohydrate metabolism and in the maintenance of the integrity of red blood cells. It protects the membranes of liver-cell mitochondria (the cell's energy factory) from damage by peroxides. Glutathione peroxidase is also involved in prostaglandin metabolism and in the killing of bacteria by white cells.

There is evidence that free-radical production increases with age. [29] Superoxide radicals are produced by the mitochondria (energy factories) of cells. The path of superoxide radicals produced by heart mitochondria were studied and the quantity of radicals produced were measured at different ages. Eighty percent of the radicals diffused into the matrix space where they were trapped by superoxide dismutase.

The remaining 20 percent of the superoxide radicals migrate across the mitochondria membrane into the cytosol (cell interior) where they react with various components of the mitochondrial membrane, such as polyunsaturated fatty acids.

Drs. Nohl and Hegner have found that the age-related increase in the formation of superoxide radicals is accompanied by an increase in the peroxide content of the mitochondria. Therefore, it is concluded that the free-radical chain reactions appear to exceed the homeostatic protection of the mitochondria in aging animals.

FIG. 4.2

Four steps in the free-radical attack on a cell

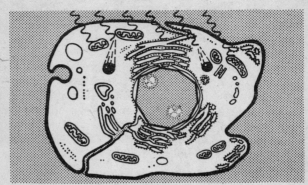

A. Radiant energy penetrating cell membrane
   initiates lipid peroxidation,
   releasing free radicals.

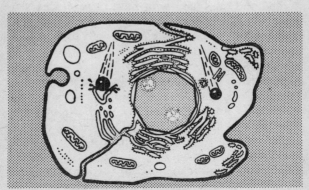

B. Free radical collides with a lysosome,
   puncturing its membranes and allowing the
   release of hydrolytic enzymes.

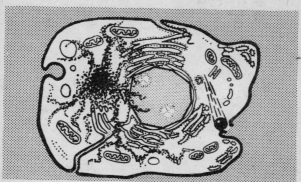

C. The hydrolysate spreads through the cell, destroying cellular components. Another free radical ruptures the cell membrane.

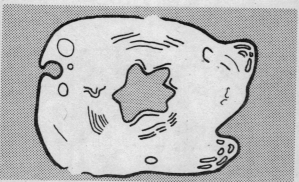

D. The end product of lipid peroxidation: a ruptured cell membrane, a distorted nucleus, and of the lysosome, only a "clinker" remains. The host has become one cell older.

Source: Reproduced with permission of *Nutrition Today* magazine. P. O. Box 1829, Annapolis, Maryland 21404. © December 1967.

FIG. 4.3 Metabolic pathways of the glutathione-Se enzyme complex

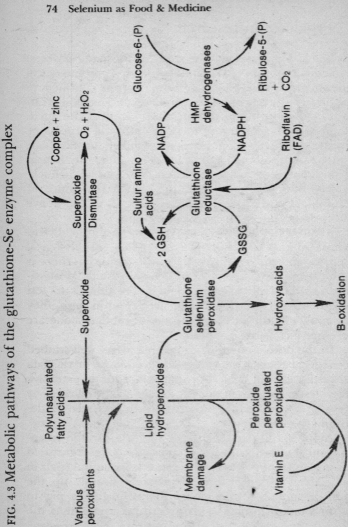

Source: W. H. Allaway, from Branda, D. and Ullrey, D. 1976, Annual Proceedings, American Association of

Problems in research

Biochemical gerontology (the study of the aging process) is still in its infancy. When I began my gerontological experiments in the 1960s, gerontology was confined to studying the effects of aging, geriatrics (the medical treatment of the aged) and the social problems of the aged. Any mentioning of trying to slow the aging process or improve longevity à la Harman, Tappel, or Strehler, brought smiles to the faces of my colleagues. Any hint at seeking rejuvenation à la Bjorksten produced loud chuckles. Just over a decade later, the National Institute of Gerontology was created to investigate all of the above concepts.

Because there are so few scientists in the field, and because of the low level of funding for gerontological research, and the lengthy experiments required, there is not an overabundance of experimental results. Most scientists in the field use small laboratory animals to test their theories. If positive results are found in mice, they may apply to humans as well. Mouse or rat data alone are therefore inconclusive at best.

In the development of drugs that affect well-studied biochemical systems, however, positive mouse or rat data can justify clinical trials. All that is needed is extensive safety testing in three species of animals, followed by safety tests in man. Then the drug can be tested to determine its effectiveness.

Aging studies are different. Because so little is understood about the subject, scientists do not feel justified in jumping from positive mouse studies to clinical trials in man. They prefer instead to progress steadily up the chain of animals—from mouse to rat to guinea pig to dog to monkey to man. This procedure strengthens the probability that a given substance will work in man and prevent the scientist from going out on a limb.

However, whether it works in man or not is not a probability but an absolute truth. Either it works or it doesn't. It could work for every animal but man, or vice versa. Only testing in man will answer the question.

In the late 1960s I calculated that it was cheaper to test possible age-retarding formulas in man than it was in monkeys. And it was obvious that I wouldn't live long enough to test my formulas which had successfully extended mouse lifespans in all the other species. The average lifespan of a dog is fifteen years and a monkey lives about sixteen years, and their maximum lifespans are thirty-five and thirty years respectively.

It seemed practical and prudent to me to break with convention and jump from successful mouse trials to clinical trials. In 1970, I presented my concepts at the 23rd Annual International meeting of the Gerontological Society in Toronto, Canada.[30,31] Several scientists added their constructive suggestions.

A brief synopsis of the theoretical aspects involved in my selection of the compound used in my successful experiments also appeared in *Chemical and Engineering News*.[32]

My approach offended a few experts in the field, unfortunately. Their criticism centered on my claim that the aging process increased the production of mis-synthesized (improperly formed) proteins. I suggested that these mis-synthesized proteins further increased protein production, caused an immunological response, and tied up nutrients.

Selenium was involved not only as an antioxidant and radiation protector, but as a means of breaking up improperly formed proteins by interaction with sulfide linkages during conformational changes. Mis-synthesized proteins are not rapidly integrated into cells due to stereochemical misfit. This concept evolved from earlier experiments by Dr. Tappel showing that selenium activated sulfhydryl enzymes.[33-34]

I defended my theory in the May 10, 1971, issue of *Chemical and Engineering News* (see Appendix B). This response provides a review of various ways in which selenium and other antioxidants prevent accelerated aging in addition to general non-specific antiradical activity. The following year a study appeared confirming protein missynthesis.[35] *Science News* commented: "Of the various theories of aging, the one that aging is due to an accumulation of defective molecules has not received a lot of attention. Yet evidence is mounting to support the hypothesis."[36] Dr. G. Roth confirmed the existence of an impaired molecular event in DNA synthesis in 1972.[37]

My views then were that aging is a combination of cell loss and the production of improper enzymes and cells. Both events are caused by free-radical reactions. My views today are that these are secondary factors that can accelerate the aging process but are not involved in the primary aging process.

My formula has proven successful in gerbils, as well as mice. In a follow-up experiment by a colleague, gerbils given the formulations lived 20 percent longer than normally nourished animals not given the formulation. Gerbils given carcinogens were protected from cancer by the formulation. The protected gerbils lived longer than normal and three times longer than non-protected gerbils.

One day I may complete the safety tests required to activate my investigational New Drug Application with the FDA[38] and proceed to clinical testing. It could mean ten to fifteen more years of healthful living to the average person.[15]

I believe the same improvement can be obtained by a moderate lifestyle with optimum nutrition. Our average lifespan should be eighty-five years, not seventy.

The evidence

The epidemiological evidence is very weak, but studies do show that lifespan correlates to soil selenium levels. Dr. David Davies has found, for example, that much of England is low in selenium, but there is one selenium-rich region where there are two to three times as many people over seventy-five years of age as in the country as a whole. In Norwich, 11 percent of the people are seventy-five years old or older, and in Upper Sheringham, 15 percent are over seventy-five. This compares to 5 percent in Britain as a whole.

Increased longevity does not mean that the aging process has been slowed. If someone examines the average lifespan in the high-selenium areas already studied for reduced cancer and heart disease rates, they most likely will see greater than average longevity.

Our best indicator at this time is the amount of age pigment (lipofuscin) present. Accumulation of age pigment increases with chronological age and correlates well with physiological age. Lipofuscin accumulation may not be a perfect indicator of physiological age as it may also vary with nutritional status. But it may also be the best age index we have. Figure 4.4 shows the accumulation of lipofuscin with age.

Dr. Tappel tested my formulation and found that it

reduced lipofuscin accumulation by 44 percent—almost cut in half—over a diet rich in vitamin E.[39] Figure 4.5 shows the fluorometric quantification of the lipofuscin present in mice on three different diets. All diets were nutritionally complete. Diet 1 was fortified with vitamin E. Diet 2 contained even more vitamin E, extra vitamin C, a synthetic antioxidant, and a sulfur-containing amino acid. Diet 3 included all of the preceding *plus selenium.*

The results imply that the lower amount of lipofuscin represents fewer free-radical reactions, thus there is less membrane damage, less DNA damage, and fewer cross-linkages. Consequently, fewer cells have been destroyed and the aging process has been slowed in comparison to less protected animals.

Patients with neuronal ceroid lipofuscinosis, a disease having accumulations of pigment in nerve cells that leads to mental retardation, visual failure, ataxia and death, are believed to have an impairment of their selenium-uptake mechanism.[40]

SUMMARY

Selenium plays an important role in reducing the damage that accelerates the aging process. A selenium deficiency can age you faster, while the optimum amount of selenium will hold your aging rate down to the basic genetic rate. The difference can mean as much as ten to fifteen years of healthy life.

❑ Selenium increases animal lifespans.
❑ Selenium reduces free-radical damage that can accelerate aging.
❑ Selenium reduces the accumulation of the age pigment.

FIG. 4.4

Accumulation of lipofuscin with age

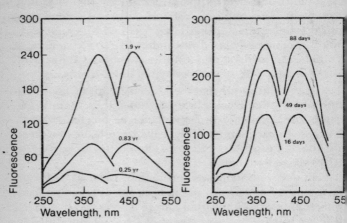

Left: Increase in extractable fluorescence in mouse testes with age.
Source: Tappel, A., Fletcher, B. and Deamer, D. 1973. *Journal of Gerontology* 28:415–24.

Right: Increase in extractable fluorescence in aging Drosophilia melanogaster.

Source: Sheldahl, J. A. and Tappel, A. 1974. *Experimental Gerontology* 9:33–41.

REFERENCES

1   Summerhayes and Franks. 1979. *Journal of the National Cancer Institute* 62:1017.

2   Baird and Birnbaum. 1979. *Chem. Biol. International* 26:245.

3   Baird and Birnbaum. 1979. *Cancer Research* 39:4752.

4   Finch, 1979. *American Journal of Medicine* 66:899.

FIG. 4.5

Fluorometric quantification of lipofuscin in relationship to three different diets

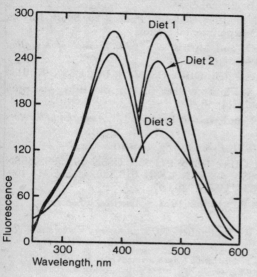

Source: Tappel, A. 1971. *Pathological Aspects of Cell Membranes*, eds. B. Trump and A. Arstilla. New York: Academic Press.

5 Harman, D. 1957. *Journal of Gerontology* 12:257–63.

6 Harman, D. 1968. *Journal of Gerontology* 23:478.

7 Harman, D. 1969. *Journal of American Geriatrics Society* 17:728.

8 Scott, M. 1962. *Nutrition Abstracts Review* 32:1.

9 Shimazu, F. and Tappel, A. 1964. *Radiation Research* 23:210–17.

10 Tappel, A. 1969. *American Journal of Clinical Nutrition*, vol. 22.

11 Bjorksten, J. 1951. Crosslinkages in protein chemistry. In *Advances in Protein Chemistry* vol. 6. eds. Anson and Edsall. New York: Academic Press.

12 Olson, O. et al. 1958. *Technical Bulletin* 20, South Dakota Agricultural Experimental Station.

13 Levander, O. and Morris, V. 1970. *Journal of Nutrition* 100(9):1111–18.

14 United States Patent Office USSN 39, 142, USSN 97, 011.

15 Passwater, R. and Welker, P. April 1971. *American Laboratory* vol. 3(4):36–40.

16 Passwater, R. and Welker, P. May 1971. *American Laboratory* 3(5): 21–25.

17 December 1971. *Prevention*, p. 104–10.

18 U.S. Patent Office USSN 271,655, USSN 398,596, USSN 481,788, USSN 593,812, USSN 613,420, USSN 718,469, USSN 806,534, USSN 930,657.

19 Passwater, R. June 1973. *American Laboratory* vol 5(6):10–22.

20 Passwater, R. September 1972. *American Laboratory* vol. 4(9):23–35.

21 Pryor, W. *Chemical and Engineering News*, June 7, 1971, p. 35.

22 Harman, D. 1968. *Journal of Gerontology* 23:476.

23 Harman, D. 1969. *Journal of American Geriatrics Society* 17:721.

24 Tappel, A. 1965. *Federation Proceedings* 24:73.

25 Tappel, A. 1967. *Biochim. Biophys. Acta* 136:402.

26 Tappel, A. 1968. *Geriatrics* 23:97.

27 Tappel, A. 1971. *Pathological Aspects of Cell Membranes*, eds. B. Trump and A. Arstila. New York: Academic Press.

28 Flohe, L. 1976. *Proceedings of the Symposium on Selenium-Tellurium in the Environment*. Pittsburgh, Pennsylvania: Industrial Health Foundation, pp. 138–157. Also see Tappel, A. 1974. *American Journal of Clinical Nutrition* 27:960–965.

29   Nohl and Hegner. 1978. *European Journal of Biochemistry* 82:562.

30   Passwater, R. October 1970. *Abstracts of the 23rd Gerontological* Society, Toronto.

31   Passwater, R. Autumn 1970. *The Gerontologist* 10(3), part II, p. 28.

32   October 26, 1970. *Chemical and Engineering News*, p. 17.

33   Caldwell, K. and Tappel, A. 1965. *Archives of Biochemistry and Biophysics* 112:196–200.

34   Dickson, R. and Tappel, A. 1969. *Archives of Biochemistry and Biophysics* 131:100–10.

35   Holliday, R. and Tarrant, G. July 1972. *Nature*.

36   *Science News*, July 29, 1972, p. 75.

37   Roth, G. *Chemical and Engineering News*, September 11, 1972, p. 22.

38   Food and Drug Administration, RF 9291 (1970).

39   Tappel, A. 1971. *Pathological Aspects of Cell Membranes*, eds. B. Trump and A. Arstilla. New York: Academic Press.

40   Sandholm, M. and Westermarch, T. 1976. *Proceedings of the Symposium on Selenium-Tellurium in the Environment.* Pittsburgh, Pennsylvania: Industrial Health Foundation, pp. 268–74.

FIVE

Arthritis

Arthritis is a multi-factorial disease for which there is no agreed upon cause or cure. There is evidence, however, that stress and/or nutritional deficiencies are causes in many people. In one case it may have been a prolonged sub-clinical vitamin C deficiency resulting in connective tissue abnormalities. In another person, a panthothenic acid (vitamin B-5) deficiency may have been involved.

Regardless of the cause of arthritis, there is strong evidence that selenium helps relieve its symptoms. Veterinarians have found that arthritic pain and swelling in animals can be substantially alleviated by treating with selenium. They use a product which contains 1000 mcg of selenium and 68 i.u. of vitamin E and report dramatic success with this combination of nutrients.[1] The cellular protection afforded by selenium is consistent with an anti-inflammatory effect that could account for reduced pain and disappearance of symptoms. This may result from the fact that the inflammatory damage in arthritis is produced in large part by free radicals. Selenium, through its role in the glutathione peroxidase enzyme, acts as a potent detoxifier of free radicals.

The anti-inflammatory property of free-radical quenchers in relieving arthritis symptoms has been well documented with superoxide dismutase (SOD). It has been reported that SOD enabled Seattle Slew to overcome equine arthritis to capture the coveted Triple Crown.[2] SOD was described in Chapter Four on the aging process as an enzyme that destroys the superoxide radical. SOD has been well tested for treatment of rheumatoid arthritis, osteoarthritis, muscular dystrophy and the side effects of radiation therapy. Research may soon begin on SOD in cancer prevention.

So far controlled clinical trials in the U.S., France, Belgium and England have found SOD more effective for arthritis than placebos and as effective as conventional gold therapy.[2] Dr. James L. Goddard, former U.S. Assistant Surgeon General labels SOD "a remarkable substance for treating both rheumatoid arthritis and osteoarthritis, which afflict more than 22 million Americans. It will relieve the suffering of millions of people."[3]

If the free-radical scavenger SOD reduces inflammation and pain, it may be that the free-radical scavenger selenium as part of glutathione peroxidase also will be effective. The limited tests conducted so far indicate that this is the case.

At the May 1980 meeting of selenium scientists, Norwegian researchers reported the beneficial results of selenium against arthritis: "In rheumatoid arthritis, it has been suggested that superoxide radicals and lipoperoxides can be generated in the tissues and accelerate the progression of the disease. Since selenium is a component of the protective enzyme glutathione peroxidase, we determined the blood levels of selenium in a group of twenty-three rheumatoid arthritis patients."[4]

The rheumatoid arthritis group did have depressed selenium levels compared to a reference group. The researchers have begun supplementation of selenium and vitamin E to the patients to assess its effects on the progress of the disease.

Another physician at the selenium conference, Dr. E. Crary of Smyrana, Georgia, had already treated patients having traumatic arthritis with selenium and the antioxidant vitamins A, C and E, successfully relieving the pain in their traumatized joints.[5]

SUMMARY

Selenium has the potential for relieving the symptoms of arthritis while stopping its progress. More studies are required to confirm the early small-scale trials.

☐ Veterinarians find selenium and vitamin E effective against arthritis.

☐ A companion free-radical scavenger, SOD, is effective against arthritis in humans.

☐ Arthritis patients have low blood selenium levels.

☐ Selenium effectively relieved arthritis in a small clinical trial.

REFERENCES

1 Hunter, H. and Lusk, G. *Bestways*, May 1976, p. 63.
2 *Family Health*, May 1980.

3 Nesi, T. *National Enquirer*, March 18, 1980, p. 56.
4 Aaseth, J., Thomassen, Y. and Munthe, E. May 1980. Second International Symposium on Selenium in Biology and Medicine, Texas Tech University, Lubbock, Texas.
5 Crary, E. May 1980. Second International Symposium on Selenium in Biology and Medicine, Texas Tech University, Lubbock, Texas.

SIX

Selenium strengthens your immune system

STRENGTHENING your immune system means increasing your resistance to disease. Selenium has improved the protective level of our immune systems by as much as twenty to thirty times.

Diseases are often classified according to our knowledge of their causes. Infectious diseases can be caused by microorganisms, such as bacteria, viruses or fungi. Other diseases may be caused by nutritional deficiency, genetic defect, injury, poison or carcinogen. A few diseases whose causes are still unknown are classified as being of "unknown etiology." Infectious diseases caused by microorganisms can be controlled by the appropriate antibacterial, anti-viral or fungicidal agents. Viral diseases can be prevented by vaccination with the appropriate vaccine.

In Chapter Two, I mentioned briefly that selenium's ability to reduce cancer risk may result in part from improved immunological destruction of newly formed tumor cells. Selenium may also improve your resistance to many diseases caused by viruses or bacteria. This encompasses a wide range of diseases, from the common cold to Legionnaire's disease and malaria. Selenium may also aid against immunologically related inflammation disease, such as arthritis.

In order to understand better how selenium protects us against so many different diseases, let's take a brief look at the immune system itself.

The immune system

The body protects itself against invasion by disease microorganisms through its ability to recognize itself and to reject and destroy everything that is non-self. The mechanism that accomplishes this is the immune system. As Dr. Robert Good of the Sloan-Kettering Institute for Cancer Research puts it, "Man lives in a sea of microorganisms; the immune system is his license to survive."[1]

The immune system is a complex network of lymphocytes (white blood cells formed in the lymph tissue), macrophages (large scavenger white blood cells), antibodies (proteins that can react with specific germs) and interferon (an anti-viral and anti-tumor compound).

The immune system produces three types of lymphocytes called T-cells, B-cells and K-cells. The T-cells have the ability to reject all foreign matter, while B-cells have the ability to produce antibodies. Relatively little is understood about the T-cells and B-cells, but less is known about the "killer" K-cells which are involved in an immune phenomenon called "antibody-dependent cell-mediated cytotoxicity."

All three specialized lymphocytes are formed from

the same basic cell of the bone marrow called a "stem cell." Figure 6.1 illustrates how stem cells are converted into T-cells, B-cells and K-cells. Stem cells are transformed into T-cells in the thymus, the small organ just under the breastbone in children that begins to atrophy. Stem cells are converted into B-cells possibly in the bone marrow or a yet-to-be-located area equivalent to an organ in chickens called the *bursa of Fabricius.*

These specialized lymphocyte cells (B-, T- and K-cells) are stored in the lymphoid tissues such as the lymph nodes under the arms, behind the ear, in the groin and in other locations.

When an invader (antigen) is detected, the immune system responds by releasing one or more of its specialized lymphocytes. Released T-cells multiply and surround the invader. Once the invader has been isolated by rings of T-cells it is chemically attacked by these defenders. The T-cells can also summon macrophages from the reticuloendothelial system to digest the invader.

The B-cells can also be released; they then produce antibodies (immunoglobulins) that stick to the invader, thereby increasing the likelihood of its ingestion by macrophages. Antibodies immunoglobin M (I_gM) and immunoglobin G (I_gG) circulate in the blood, while immunoglobin A (I_gA) circulates in saliva and fluids that bathe mucous membranes. At least one other antibody, immunoglobin E (I_gE) exists, but little is known about it.

Antibodies are tailor-made to specifically lock onto each of the millions of different microorganisms that may invade a person. Macrophages cleanse the blood and lymph of foreign particulate matter and also produce interferon, the body's anti-viral compound. Dr. Norbert Roberts of the University of Rochester Medical School notes, "In addition, macrophages' production of interferon may be

FIG. 6.1
Schematic showing the major components of the immune system

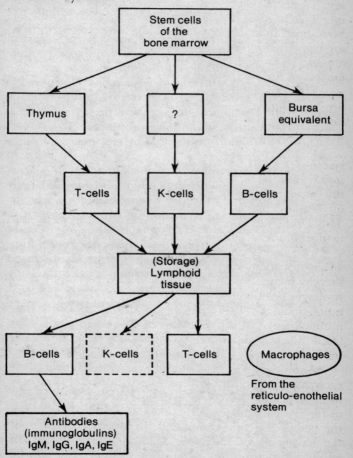

one mechanism of their extensive *antitumor* activity, since interferon can render macrophages tumorcidal and can inhibit the growth of tumor cells directly. Interferon has not only antiviral activity, but it also has cell-growth inhibitory activity."[2]

In 1973, Dr. Douglas Frost of Brattleboro, Vermont, pointed out that selenium stimulated ubiquinone production which in turn stimulated the immune system.[3]

Mark McCarty, Director of Research of Nutrition 21, speculates that selenium has a direct role in macrophage production.[4] He points out that the efficiency of macrophage production in immunological function is relatively labile (unstable) and is influenced by metabolic and pharmacological factors.

Mark McCarty notes that the efficiency of macrophage function is energy-dependent and hinges largely upon mitochondrial function. The mitochondria are particularly susceptible to structural and functional disruption by free radicals. Macrophage mitochondria are particularly subject to free-radical attack due to the compounds normally produced in macrophages. As discussed in Chapter Four and elsewhere, selenium helps the body to dissipate free radicals before they can damage key body components. Appendix C presents the details of McCarty's hypothesis.

The evidence

In 1972, Soviet researcher Dr. T. F. Berenshtein discovered that supplements of selenium plus vitamin E produced more antibodies when a vaccine was given than when the vaccine was given alone. Dr. Berenshtein found that selenium alone was less effective than the combination, and vitamin E alone had no effect on antibody formation. Dr. Berenshtein measured the antibodies produced during the immunization of rabbits with typhoid vaccine.[5]

In 1973, studies indicated a similar stimulation of the immune response in mice by selenium. Drs. John Martin and Julian Spallholz and their colleagues at Colorado State University found that dietary supplementation with selenium at levels above those recommended as nutritional requirements enhanced the primary immune responses.[6] (Drs. Spallholz and Martin are now at Texas Tech University.) The Colorado State University researchers measured the number of antibody-forming cells and the levels of antibodies in sheep. The stimulatory effect of selenium was independent of vitamin E levels.

Selenium promoted the increased number of immunoglobin-M-producing cells and immunoglobin-M antibodies. Diets at 0.7 parts per million and 2.8 parts per million produced respectively 7-fold and 30-fold more antibodies.

In a later symposium, the researchers also reported enhanced secondary antibody response to tetanus toxoid. Their experiments showed selenium enhanced the immune response in guinea pigs. They attributed selenium's role in stimulating the immune response to provoked inflammation, increased ubiquinone production and depressed cyclic AMP (adenosine monophosphate, a nucleotide) formation.[7] They concluded, "Clearly, dietary selenium enhances both primary and secondary immune response. There is an increased production of both immunoglobin-M and immunoglobin-G."

In 1978 experiments with dogs confirmed the earlier observations (in mice, sheep and guinea pigs) that selenium enhanced the immune response. Drs. B. Sheffy and R. Schultz found that antibody production in response to canine distemper-infectious hepatitis virus vaccine was dependent upon both selenium and vitamin E status of the dog.[8]

Selenium deficiency has been implicated in the once baffling illness called Legionnaires' disease that was first noticed at a Philadelphia convention in 1976. Dr. J. E. McDade and his colleagues identified the bacterium that causes Legionnaires' Disease in 1977.

The blood from seventeen patients was analyzed for several minerals. All seventeen patients had low blood levels of selenium, while there were no significant findings concerning the other minerals.[9]

Few people have since been diagnosed as having Legionnaires' disease although many have been exposed to the bacterium. Could it be that only those who are most selenium-deficient can contract the disease?

In 1980, scientists started taking advantage of the immune enhancement property of selenium and used selenium supplements to strengthen the protective effect of a malaria vaccine.[10]

SUMMARY

Experimental data and epidemiological surveys support the findings that selenium deficiency weakens the immune response and increases the susceptibility to infection and cancer. Optimal levels of dietary selenium stimulate the immune response to provide superior protection against disease.

☐   Selenium has been shown to improve the immune response in several animal species.

☐   Selenium deficiency decreases the immune response.

☐   People in low-selenium areas have more cancer than people in high-selenium areas.

REFERENCES

1 *Time*, March 19, 1973, p. 67.

2 Roberts, N. March 4, 1980. *National Institute of Health Record*, p. 7.

3 Frost, D. and Van Poucke, R. 1973. In *Trace Substances in Environmental Health* ed. D. Hemphill. Columbia: Univ. of Missouri.

4 McCarty, M. December 4, 1979. Private communication.

5 Berenshtein, T. 1972. Zdravookh. Belorussia 18(10):34–6. C.A. 78:24162.

6 Spallholz, J. et al. 1973. *Proceedings of the Society of Experimental Biological Medicine* 143:685–98.

7 Martin, J. and Spallholz, J. 1976. *Proceedings of the Symposium on Selenium-Tellurium in the Environment*. Pittsburgh, Pennsylvania: Industrial Health Foundation, pp. 204–25.

8 Sheffy, B. and Schultz, R. 1978. *Cornell Veterinarian* 68 (Suppl. 7):48–61.

9 Chen, J. and Anderson, J. 1979. *Science* 206:1426–27.

10 Desowitz, R. and Barnwell, J. 1980. *Infection and Immunity* 27(1):87–89.

SEVEN

The growing danger of radiation

T HE accident at the nuclear energy plant on Three Mile Island near Harrisburg, Pennsylvania in 1979 has awakened many people to the dangers of unexpected radiation. A nearby reactor is not the only problem, as nuclear wastes are being transported by train and truck.

Furthermore, medical diagnoses and treatment can subject you to dangerous levels of radiation be it X ray, isotope screening, or cobalt therapy. And even background radiation can cause birth defects.

What can you do in an emergency to lessen the damage from radiation? Your radiation defense "first aid" kit should include kelp or potassium iodine tablets, superoxide dismutase tablets and antioxidants including vitamins A, C, E, and the mineral selenium. In fact, selenium compounds may offer the greatest protection against radiation with superoxide dismutase (SOD) a close second. As mentioned in Chapter Five on arthritis, SOD has been shown effective against radiation in clinical trials. SOD is a partner to selenium in quenching free radicals. This lends verification to the theory that free-radical scavengers protect against radiation-induced inflammation.

In 1964, Dr. Tappel pointed out that the evidence suggests that the ideal radiation protector is a molecule that can release and accept electrons and hydrogen atoms easily, without itself becoming dissociated.[1] Dr. Tappel noted that sulfur-amino acids were thought to be excellent radiation protectors and he theorized that selenium-containing amino acids would be better: "The ionization potential and bond energy of selenium compounds are smaller than that of sulfur compounds."

That is just what his experiments showed. Selenoamino acids were powerful radiation protectors. Dr. Tappel outlined three free-radical scavenger and repair mechanisms and two non-radical mechanisms, by which seleno-amino acids protect against radiation.

In 1969, Dr. G. Colombetti of Pisa, Italy, confirmed Dr. Tappel's finding and also established that the selenoamino acids were themselves very resistant to radiation.[2,3]

In 1979, Dr. Roberto Badiello of the University of Bologna, Italy, and Dr. Martin Fielden of the Institute of Cancer Research in Sutton, England, further tested the protective effect of selenium compounds.[4] They expanded the earlier research of Dr. Badiello that found selenourea to be an excellent protector against radiation.[5,6] Through their measurement of the action of the selenium compound on the primary free radicals that form when water is radiated, we now have a better understanding of the value of organic selenium compounds.

In the early 1970s, research substantiated the concept that natural selenium amino acid and protein compounds are tremendous protectors against radiation.[7,8]

Dominick Bosco reported in *Prevention* (August 1976) that Dr. Julian Aleksandrowicz, director of the Hematological Clinic of the Institute of Internal Medicine, Medical Academy of Krakow, Poland, found selenium extremely protective against irradiation by gamma rays. Dr. Aleksandrowicz irradiated two groups of ten mice. Of the ten mice supplemented with selenium prior to total body irradiation, eight survived. Only one of the ten not given selenium prior to irradiation survived more than two weeks.

SUMMARY

Daily exposure to background radiation produces enough damage to accelerate our aging process and cause cancer and birth defects. Your defense arsenal should include organic selenium compounds, which are proven protectors against radiation.

REFERENCES

1   Shimazu, F. and Tappel, A. 1964. *Radiation Research* 23:210–17.

2   Colombetti, G. 1969. *Studia Biophysica* 18(1):51–55.

3   Colombetti, G. and Falcone, G. 1969. *Lettere Nuovo Cimento* 2:127–30.

4 Badiello, R. and Fielden, M. 1970. *International Journal of Radiation Biology* 17(1):1–14.

5 Badiello, R. et al. 1967. *Med. Nucl. Radiobiol. Latina* 10:57.

6 Breccia, A. et al. 1969. *Radiation Research* 38:483.

7 Colombetti, G. and Munti, S. 1971. *Proceedings of European Biophysics Conference* 2:45–53.

8 Badiello, R. et al. 1971. *International Journal of Radiation Biology* 20(1):61–68.

EIGHT

Cystic fibrosis

Cystic fibrosis is considered a disorder of the exocrine glands which overproduce a thick mucus that clogs the lungs and digestive system. The vast majority of researchers believe that cystic fibrosis is caused by a genetic defect. One researcher, Dr. Joel Wallach, has considerable data suggesting that cystic fibrosis is due to a selenium deficiency in the mother. Which view is correct? Or could both be partially correct?

We have learned that salt causes high blood pressure only in those that are genetically susceptible. Perhaps cystic fibrosis is transmitted only by mothers who are both carriers of the defective gene and selenium deficient. If this is the case—and it is only a hypothesis—then selenium could prevent cystic fibrosis.

An unconfirmed, but certainly plausible, report in the June 1979 Proceedings of the National Academy of Sciences suggests that the alleged genetic defect involves the enzyme NADH dehydrogenase. Dr. Burton Shapiro of the University of Minnesota first noticed that cystic fibrosis patients and carriers have excessive calcium between their cells. Dr. Shapiro then traced this to the energy production system in the mitochondria (see Chapter Five). Later Dr. Shapiro and his colleagues traced this problem to a difference in the enzyme NADH dehydrogenase.

Although cystic fibrosis occurs as often as one in 2,500 births among Caucasians of Middle European extraction (less in other ethnic groups), research has been hampered by the lack of a suitable animal model. However, a male rhesus monkey was born in May 1977 that was unequivocally confirmed upon autopsy as having cystic fibrosis.[1]

Shortly thereafter, five more monkeys—biologically unrelated to the first but born in the same compound—were confirmed as having cystic fibrosis.

Dr. Wallach, an expert in comparative pathology, also noted that the autopsied organs looked exactly like organs from selenium- and zinc-deficient animals (see Figure 8.1). He then checked for possible selenium and zinc deficiencies in mothers having children with cystic fibrosis.

A survey of 15 mothers with cystic fibrosis children being treated at the Grady Memorial Hospital (Atlanta) revealed a total of 48 pregnancies, many of which were stormy and complicated. In addition to the cystic fibrosis children, there were ten miscarriages and multiple birth defects were reported in one infant. Two of the mothers in this group were treated during pregnancy to prevent abortion. All of the mothers in this group were employed

FIG. 8.1

Organ changes in cystic fibrosis related to selenium deficiency

	Pancreatic atrophy	Pulmonary disease	Hepatic necrosis/fibrosis	Malabsorption syndrome	Subcutaneous edema	Muscular fibrosis	Cardiac necrosis or fibrosis	Urogenital malformations or necrosis
Human Cystic Fibrosis	+	+	+	+	+	+	+	+
Monkey Cystic Fibrosis	+	+	+	+	+	+	+	+
Rat	+	+	+	+	+	+	+	+
Mouse	+	?	+	+	−	+	+	+
Chick	+	?	?	+	+	+	+	?
Sheep	+	+	?	+	?	+	+	+
Pig	?	+	+	+	?	+	+	+

Organ changes compatible with those seen in human cystic fibrosis have been congenitally and neonatally induced in a wide variety of animals with either maternal or neonatal selenium deficiency.

Source: Wallach, J. May 25–26, 1978. Paper presented at the Workshop on Model Systems for the Study of Cystic Fibrosis, Bethesda, Maryland.

in busy responsible jobs and had less than ideal meal programs and dietary habits. Eight pregnancies resulted in significant hair loss or notable change in hair character, features associated with selenium and zinc deficiencies in laboratory animals. Three pregnancies were associated with anemia, two with pre-eclampsia, two with maternal edema, and one mother was treated for a kidney infection during pregnancy. A total of 27 (56 percent) of the 48 pregnancies produced cystic fibrosis, miscarriages or congential birth defects. This data is consistent with and further supports the proposed theory of an environmental etiology for cystic fibrosis. It is also of interest that identical twins affected with cystic fibrosis usually present with one twin more severely affected than the other, indicating fetal competition for an essential nutrient or nutrients that are present in limited and less than optimal amounts.[2]

Dr. Wallach next examined cystic fibrostic children for selenium levels. He found the value very low. The secondary general malnutrition of the disease itself could not account for such a low selenium level so early in life.

Cystic fibrosis also occurs more frequently in low-selenium areas in the United States and the world as a whole.

The diet fed the monkeys was found to contain excessive polyunsaturated oils added to improve the gloss of their fur. This increased the consumption of antioxidants, thus depleting vitamin E and selenium from the mothers. Blood analyses taken routinely during their pregnancies showed very high creatine phosphokinase levels which is a sign of vitamin E deficiency.

In a later report, Dr. Wallach expanded his earlier study.

A study of 120 families with one or more cystic fibrosis children was conducted by voluntary telephone contact or office interview to accumulate data which would support or refute the environmental-selenium deficiency theory of cystic fibrosis etiology. A total of 328 pregnancies were identified of which 168 were cystic fibrosis children (51.2 percent), thirty-four were miscarriages (10.36 percent), four were stillborn (1.2 percent), two were S.I.D.s (0.6 percent), two had hyaline membrane disease (0.6 percent), six had congenital anomalies (1.8 percent) and 114 were normal (34.75 percent). The families interviewed represented thirty-three states and four Canadian provinces. The cystic fibrosis patients ranged in age from three months to thirty-seven years of age.

Forty-two cystic fibrosis cases were diagnosed at birth, forty-three were apparently normal at birth and developed the disease before nine months of age, thirty cystic fibrosis cases were apparently normal at birth and developed the disease between nine months and three years of age, twenty-five cases were apparently normal at birth and developed the disease between three years and ten years of age, ten cases were apparently normal at birth and developed the disease between ten and twenty years of age, three cases were apparently normal at birth and developed the disease between twenty and thirty years of age, twelve cases developed the disease at unknown ages. Thirty cases investigated had died prior to entering the corrective program.

One hundred and fifty-one cases were associated with prenatal, lactation or prediagnosis diets rich in polyunsaturated fatty acids (vegetable oils or fish oils) and low in the guidepost foods potentially rich in selenium (eggs, whole milk, liver, kidney or unpolished rice). Seventy-one cystic fibrosis infants were fed milk replacers having a vegetable oil base, three cystic fibrosis cases received cod liver oil in their formula or diet and forty-two cystic fibrosis infants were initially breastfed prior to being fed commercial formulas. One hundred and six parents of cystic fibrosis children had overt symptoms consistent with selenium deficiency (gall bladder diseases, anemia, heart disease, alopecia, achromotrichia, exudative diathasis, infertility and miscarriages).

Selenium analysis of an acute cystic fibrosis case revealed low whole blood selenium; selenium analysis of stabilized cystic fibrosis cases revealed values in the low to middle normal levels.

Post mortem tissue analysis of two infant cystic fibrosis cases outside this study revealed selenium levels of 1/10 the expected normal values. One mother breastfeeding a three-month-old cystic fibrosis infant had a hair selenium value of 25 percent of the normal expected values.

Fifty cystic fibrosis patients ranging in age from three months to thirty-seven years of age have participated in a diet program based on a balanced diet, free of vegetable oils and with selenium supplementation ranging from 25 to 300 mcg per day. All participating individuals have displayed markedly improved clinical status (normal B.M. pattern, reduced lung mucus, increased energy and strength, improved skin and hair quality, rearchitecturing of barrel chests, reduced finger and toe clubbing, weight gain, increased resistance to infections and improvement in voice quality).

The natural history and patient profile of the predominantly perinatal disease complex known as cystic fibrosis fits that of an acquired environmental disease that can be produced by a deficiency of selenium, zinc and riboflavin and can be aggravated and precipitated by a low vitamin E diet that is rich in polyunsaturated fatty acids.[3]

Dr. Wallach's discussion is supported by 198 references before reaching its conclusion. "The direct and indirect evidence associated with the index cystic fibrosis rhesus and cystic fibrosis human strongly suggest that cystic fibrosis is an environmentally induced disease rather than genetically transmitted as a simple Mendalian auto-somal recessive disease.

"The metabolic pathways associated with biological membrane protection involves selenium, vitamin E, zinc, copper and riboflavin; membrane injury produced by peroxidation of long chain polyunsaturated fatty acids and the known changes produced in the pancreas, respiratory system, gastrointestinal tract, heart and geni-tourinary system by deficiencies of these micronutrients are consistent with the pathological changes and clinical disease associated with human cystic fibrosis."[1]

Dr. Wallach's observations imply that there is more hope for cystic fibrosis patients who have not yet had severe pancreatic fibrosis. This is reversible in animals treated with selenium.[4]

Dr. Wallach points out that while cystic fibrosis can be explained by a selenium deficiency in the mother, most infants born with cystic fibrosis do not immediately exhibit pancreatic disease.

Dr. Wallach's findings are the result of intensive and careful study, well-documented by the accepted scientific findings of others. His theory is in harmony with the facts, and thus deserves follow-up. Yet this has not been the case.

Rather than further test Dr. Wallach's theory, the establishment has chosen to squelch Dr. Wallach's research. Within twenty-four hours of announcing his findings to his superiors, Dr. Wallach was dismissed from the Yerkes Regional Primate Center in Atlanta. He is at this writing an Associate Professor in the Department of Nutrition of the National College of Naturopathic Medicine in Portland, Oregon.

FIG. 8.2

Maternal selenium levels related to congenital pancreatic cystic fibrosis

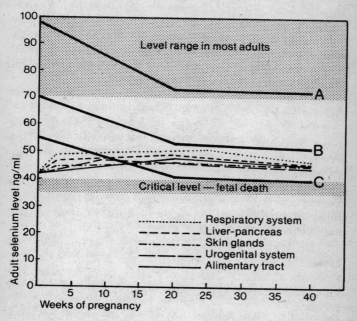

Maternal serum selenium levels are overlaid on graphs showing time frames of human fetal organogenesis. High line (A) and middle line (B) represent high and low normals respectively. Hypothetically low maternal serum selenium (line C) can explain the production of congenital pancreatic cystic fibrosis and the contribution to the production of cystic fibrosis lesions in other organ systems.

Source: Wallach, J. May 25–26, 1978. Paper presented at the Workshop on Model Systems for the Study of Cystic Fibrosis, Bethesda, Maryland.

Treatment

Dr. Wallach has developed a therapy for cystic fibrosis that has been very effective for many people. Dr. Wallach recommends that cystic fibrosis patients should improve their nutrition by eating a green salad daily, lean meats and liver four times a week, eggs for breakfast daily, avoid wheat and replace with rice and potatoes, eat gelatin daily, and drink milk with every meal.

Dietary supplements of selenium, zinc, copper, riboflavin (vitamin B-2), vitamin E and vitamin K should be taken daily.

The ABC television network discussed Dr. Wallach's therapy on the December 13, 1979 segment of "20/20." ABC news interviewed a toxicologist to counter the claims of Dr. Wallach. A toxicologist is not an expert in cystic fibrosis but is expert in what happens when you get too much of anything.

The toxicologist implied that the death of one of the 300 patients was the result of following Dr. Wallach's suggested nutrient regimen. The child was said to have received 25 micrograms of selenium daily. There is not the slightest possibility that selenium could be toxic at that level, as the 1979 Food and Nutrition Board of the National Academy of Sciences' recommendation is a daily intake of 20 to 60 micrograms per day for infants six months to one year of age.

Some selenium experts recommend that people take even more than the official RDAs. Children in some parts of the world eat 300 to 700 micrograms of selenium daily and are in excellent health. Dosage and toxicity will be discussed in later chapters, but the point is that uninformed members of the establishment are quick to strike out and

attempt to defame anyone suggesting that nutrition prevents or helps control any disease.

The toxicologist repeated that "selenium is toxic." A
scare tactic. So are iron, vitamin A, vitamin D, zinc, salt,
water and everything else—when too much is taken.
Apparently, the toxicologist did not know of the nutritional
essentiality of selenium—or the biochemistry of selenium
in the prevention of fibrotic damage.

The success of Dr. Wallach's therapy has been reported in the popular nutritional literature.[5] The nutrients
have helped patients from a few months of age to thirty-
four years, but I will report only one here as an example.

A two-day-old child in Gering, Nebraska, was diagnosed as having cystic fibrosis. This diagnosis was confirmed in several clinics over the next five months. The
baby girl would just lie still in bed as it was difficult for
her to even breathe. Her weight cycled between 4 and 7
pounds.

After five months on Dr. Wallach's therapy, however, she weighed 12.3 pounds, her lungs were *completely
cleared*, and she was rolling over and playing with toys.
She was again tested for cystic fibrosis, with a negative
result.

During the second International Symposium on Selenium in Biology and Medicine held at Texas Tech University in Lubbock, Texas, in May 1980, a group of
researchers from the Children's Hospital at Stanford, California, headed by Dr. Ricardo Castillo, reported:

> The Pediatric Pulmonary Service at the Children's
> Hospital at Stanford follows approximately 250 patients
> with CF in all degrees of illness severity. Serum Se levels
> were measured in 25 randomly selected patients from this
> population. Blood was collected from patients in the fasting

state and placed in trace element-free containers. The serum was seperated and stored at $-20°C$ until shipped to the USDA Institute of Nutrition at Beltsville, MD., for analysis.

The mean serum Se level for our patients sampled was .1054 ± .0312 meq/ml with a range of .0416 mcg/ml to .1495 mcg/ml. No correlations with age, Schwachman Index, growth retardation, severity of malabsorption, caloric intake, delayed sexual development, or serum zinc levels were found.

Although our CF patients had serum Se levels below what might be considered a typical U.S. value of 0.150 mcg/ml, all but one patient had levels above the normal value for Otago, New Zealand, which is 0.043 mcg/ml. Because there is no evidence of increased incidence or severity of CF in New Zealand, our conclusion is that our CF patients are not Se deficient as assessed by serum Se levels. Based upon these data, there is therefore no indication for Se supplementation in patients with CF. The relative difference in the typical U.S. value for serum Se and those values found in our patients must be interpreted in light of the ages of the patients and regional differences. We are therefore currently in the process of gathering and analyzing serum from age-matched controls from the San Francisco Bay area which should allow more accurate interpretation of this difference.

I believe it is unfair to compare populations of all the U.S. and Central Europe to that of pollution-free New Zealand, unless further evidence justifies this. Their essentially cadmium-and-ozone-free environment also lessens their need for selenium. New Zealanders eat a large amount of vitamin E-rich vegetation. Vitamin E spares selenium. Also, we need to know more about the actual rate of birth defects in this population.

Dr. Roy F. Goddard, president of the Children's Lung

Association, has had excellent results by supplementing the diets of cystic fibrosis children with selenium. Paul Einstein quoted Dr. Goddard in the July 15 *National Enquirer:* "I think this is one of the biggest breakthroughs since the discovery of the disease."

Dr. Goddard reported that he had twelve patients on the diet at his CF clinic in Alburquerque, New Mexico: they consume foods rich in trace minerals, such as eggs, milk, liver and unpolished rice, plus supplements of selenium, zinc and copper. The diet avoids foods high in polyunsaturated fats.

Dr. Goddard commented, "They're doing great. I'm not saying they're cured, but I see weight gain and changes in the gastrointestinal tract. A couple have improved their pulmonary situation."

Mr. Epstein reported similar results by Dr. Robert Mckay, director of the Cystic Fibrosis Clinic at the University of Miami School of Medicine. Dr. Mckay has twenty-eight patients on the diet. Virtually all had improvements in their bowel patterns and in their abdominal symptomatology within four months on the diet.

Researchers at the National Institutes of Health, led by Dr. Philip Farrell, have determined that cystic fibrosis patients should be given vitamin E.[6] As Dr. Farrell notes, "Taking vitamin E isn't going to cure cystic fibrosis, but it's likely that a vitamin E supplement patient will have an edge in fighting cystic fibrosis."

Why not take vitamin E plus selenium?

SUMMARY

Selenium should be further investigated for a possible protective and palliative effect against cystic fibrosis. Cystic fibrosis patients should be sure to get their optimal amounts of selenium.

☐   Cystic fibrosis resembles diseases in selenium-deficient animals.

☐   Dr. Wallach's therapy has aided many patients.

REFERENCES

1   Wallach, J. May 25–26, 1978. Paper presented at the Workshop on Model Systems for the Study of Cystic Fibrosis, Bethesda, Maryland.

2   Caplan, D. 1978. Private communication to Dr. J. Wallach.

3   Wallach, J. and Garmaise, B. June 1978. *Proceedings of the Thirteenth Annual Conference on Trace Substances in Environmental Health*, University of Missouri, pp. 469–76.

4   Gries, C. and Scott, M. 1972. *Journal of Nutrition* 102:1287–96.

5   Shaw, L. *Prevention*, October 1979, pp. 122–30.

6   Farrel, P. et al. July 1977. *Journal of Clinical Investigation* 62.

NINE

Muscular dystrophy

I T is fairly common knowledge among nutritional researchers that selenium prevents *nutritional* muscular dystrophy in animals. Most of these same scientists will quickly point out that this is vastly different from human muscular dystrophy.

There is evidence however, that both nutritional muscular dystrophy and human muscular dystrophy are helped by selenium and vitamin E.

Previously research with muscular dystrophy in animals focused on vitamin E alone, and attempts to correlate vitamin E status with human muscular dystrophy were unconvincing. Is there better correlation between the dual deficiencies of selenium and vitamin E? Let's look at some of the animal research.

In 1962, a Cornell University group demonstrated that selenium and vitamin E could prevent muscular dystrophy in lambs.[1] The researchers conducted experiments over five years wherein they fed selenium-and-vitamin-E-deficient diets to pregnant ewes until one month prior to lambing. The ewes and their lambs were then randomly assigned to experimental diets until the lambs were two months old.

The selenium-and-vitamin-E-deficient diets of mixed hay and raw kidney beans caused 24 to 60 percent of the lambs to become dystrophic. Supplementation of the ewe's diet with either vitamin E or selenium resulted in a marked reduction in the incidence of dystrophy in the lambs but never completely prevented it at the levels fed. When both selenium and vitamin E were given the ewes, dystrophy was completely prevented in the lambs.

Dystrophy was also completely prevented in the lambs by giving either selenium or vitamin E directly to them.

In 1964, Drs. W. H. Allaway and J. F. Hodgson observed that the selenium content of forages correlated with the geographic distribution of muscular dystrophy in livestock.[2]

A decade later a research group at Purdue University determined that weanling pigs fed a selenium-deficient diet had selective destruction of Type I skeletal muscle fibers.[3] Those pigs had lesions of skeletal myodegeneration, mulberry heart disease and hepatosis dietetica.

Probably the closest animal model to the human is the rhesus monkey. Muscular dystrophy develops in rhesus monkeys fed a vitamin E-deficient diet. The dystrophy rapidly progresses and invariably leads to death unless vitamin E is given.[4]

Dr. James Dinning at the University of Arkansas and his colleagues found that Coenzyme Q (ubiquinone) brought monkeys which were near death from dystrophy into rapid remission. (In Chapter Three on heart disease, the role of selenium in the production of Coenzyme Q was discussed.) The researchers commented: "That Coenzyme Q may show effects in man not seen with vitamin E requires new medical research which may be further justified by these new data on the response of the dystrophic monkeys."[5]

Dr. Karl Folkers was one of the researchers in that group. Dr. Folkers is now the director of the Institute for Biomedical Research at the University of Texas in Austin. His research with Coenzyme Q and muscular dystrophy was described in *Chemical and Engineering News* when he was awarded the International Robert A. Welch Award in Chemistry.

Following several years' work with genetically dystrophic mice during which treatment using Coenzyme Q was successful in prolonging the mice's lives, a strong shift to human disease was made. Early results of seven years' clinical studies were erratic, probably because patients tested had very advanced and irreversible muscular dystrophy. Many variables affected the tests.

As a result, Dr. Folkers now insists on treating boys in very early stages of the disease. Preclinical dystrophic boys and infants can appear rather normal, Dr. Folkers says, but their blood shows highly abnormal biochemistry.

The tests have had periods of controls, treatment, and placebo. Since monitoring such therapy takes years, microtechniques are used to analyze a few drops of blood for the most critical enzyme, creatine phosphokinase. Two young boys, particularly, have benefited in clinical trials, Dr. Folkers says, with both showing a reduction of creatine phosphokinase, indicating that they have biochemically benefited from coenzyme Q.[6]

Dr. M. Hidiroglou found the selenium and coenzyme Q content of the tissues of calves and rabbits affected with muscular dystrophy seem to confirm a biosynthetic relationship between selenium and Coenzyme Q.[7]

In a conversation about vitamin E, with Wayne Martin, author of *Medical Heroes and Heretics*, the subject turned to the muscular dystrophy in a friend of Wayne Martin's. I pointed out Dr. Folker's success with Coenzyme Q and the selenium as well as vitamin E was required. Wayne Martin tells of his friend's success with the stimulation of his own Coenzyme Q with selenium and vitamin E:

"Mr. S. went to the Cleveland Clinic in July of this year (1974) where blood tests showed creatine phosphokinase (CPK) of 610, and biopsy confirmed late stage muscular dystrophy. It was incurable, he was told to go home and bother us no more.

"Mr. S. had been taking 400 i.u. of vitamin E and 500 mg of C a day. For many years I have felt that muscular dystrophy is due to a lack of ubiquinone, that vitamin E is just not getting ubiquinone produced. I explained all this to Mr. S. and he went on 2000 i.u. of vitamin E and 3000 mg of vitamin C a day. Also he went on a diet rich in cottage cheese and tuna fish to supply selenium. With this his triglycerides dropped to 68 from 130. His total cholesterol dropped from 240 to 186 and his CPK dropped from 610 to 140. Most labs consider CPK of 140 as being borderline with respect to muscular dystrophy."[8]

Perhaps the nutritional muscular dystrophy in animals is different than the dystrophy in humans—but maybe it's not. At least some patients are responding to selenium plus vitamin E or Coenzyme Q therapy.

SUMMARY

Preliminary evidence suggests muscular dystrophy may respond or be prevented by a diet rich in selenium and vitamin E. Suggested links to vitamin E alone have been discounted, leading to the belief that animal and human dystrophies are different. That concept may change with more research on the selenium plus vitamin E link.

☐ Selenium plus vitamin E deficiencies cause dystrophy in animals.

☐ Replenishment of the two nutrients prevents and cures the disease.

☐ Coenzyme Q cured monkeys of muscular dystrophy. Selenium is needed for the production of Coenzyme Q.

☐ Dystrophic humans have responded to Coenzyme Q or selenium plus vitamin E.

REFERENCES

1 Hogue, D. et al. 1962. *Journal of Animal Science* 21:25–29.

2 Allaway, W. and Hodgson, J. 1964. *Journal of Animal Science* 23:271.

3 Ruth, G. et al. 1974. *American Journal of Veterinarian Research* 35(2):237–44.

4 Dinning, J. and Day, P. 1957. *Journal of Experimental Medicine* 105:395.

5 Dinning, J. et al. May 20, 1962. *Journal of Biological Chemistry* 84:2007–8.

6 *Chemical and Engineering News,* April 30, 1973, p.18.

7 Hidiroglou, M. et al. 1967. *Canadian Journal of Animal Science* 45:568.

8 Martin, W. October 4, 1974. Personal communication.

TEN

Cataract

CATARACT is the leading cause of blindness in adults. Although most prevalent among the elderly, it can occur in infancy as well.

A cataract involves a region of the eye lens that scatters light falling on it. The proteins in the normal lens are uniformly distributed within the fibers, hence the lens is transparent. Cataracts form where there is a change in protein distribution or localized concentration. This protein change can be brought about by oxidation.[1]

The eye lens has no direct blood supply. The lens depends upon a surrounding fluid—the aqueous humor—to transport nutrients to it. Proteins and nucleic acids are formed within the lens.

Diabetes, severe stress and malnutrition have been linked to cataract. Recent research has linked antioxidant deficiencies—especially vitamin E and selenium—to cataract. All of the preceding may cause cataract, but antioxidant deficiency may account for the largest share.

118

Drs. Margaret Garner and Abraham Spector of Columbia University College of Physicians and Surgeons in New York City have found the lenses of the young have no oxidation of their proteins. There is some oxidation in lens proteins in older healthy people, and dramatic oxidation of lens proteins of cataract patients. The researchers claim their data "clearly support the viewpoint that extensive oxidation of lens proteins occurs with cataract and that it begins at the lens fiber membrane."[1]

Actually, oxidation of lens protein had been suggested long ago, but this concept had been challenged.[2]

Glutathione peroxidase (a selenium containing enzyme) is known to catalyze the destruction of oxidizers such as hydrogen peroxide and lipid peroxides, as discussed in several earlier chapters.[3] Adequate dietary selenium is necessary to maintain glutathione peroxidase activity in the lens.

Prolonged selenium deficiency in rats produces cataracts.[4] Dr. William Hoekstra's group at the University of Wisconsin suggests, "It seems probable then, that in lens, vitamin E may function to prevent oxidative damage to sensitive membrane lipids by decreasing lipid hydroperoxide formation, selenium may function via glutathione peroxidase to prevent oxidative damage by destruction of hydrogen peroxide and lipid hydroperoxides when they are formed, and sulfur-containing amino acids may function as precursors of glutathione which acts as a substrate for glutathione peroxidase."[5]

The researchers also noted that the selenium content of cataract lens was only one-sixth of normal.

Dr. Donald Atkinson of San Antonio, Texas linked poor nutrition to cataract. He wrote in *The Eye, Ear, Nose and Throat Monthly*:

At the present time, I have under treatment more

than 450 patients with incipient (beginning) cataract. A
number of these have remained incipient over a period of
eleven years. I have possibly seen in more than thirty years
of practice of ophthalmology as many cataracts as any
ophthalmologist in my section of the country. Formerly,
nearly all these cases soon went on to maturation and
operation; at this time, only a limited number under ob-
servation and treatment have had to submit to surgery.

I believe, then, that in a larger percentage of cases
than most surgeons have realized, cataract is a preventable
disease.[6]

Dr. Atkinson was using one gram of vitamin C and
200,000 units of vitamin A in his therapy. This was in the
early 1950s—before selenium supplements and widespread
use of vitamin E.

The premier nutritionist, Dr. Roger Williams, founder
of the Clayton Foundation Biochemical Institute at the
University of Texas, found that every group of rats fed a
less than adequate diet developed cataracts.[7]

Dr. Williams also noted that giving rats with cataracts
good diets leads to improvement. In his experiments sixteen of twenty-six cataracts showed 40 to 80 percent disappearance.

Dr. Williams remarks, "Although much yet remains
to be learned about the details of cataract formation,
there is enough evidence now to indicate that faulty metabolism or metabolism inadequate for adapting to stresses
is a major factor—and evidence too, that not some panacea nutrient but rather *a balanced team of nutrients has
potential for preventing cataract formation and perhaps even for
helping in the treatment of some existing but not far advanced
cataracts.*"[8]

Many people are living in selenium-deficient areas,
thus they have an unbalanced group of nutrients. It may

be that selenium supplementation may balance their team of nutrients.

Studies have implicated inorganic sodium selenite in causing cataracts, perhaps because this inorganic form of selenium oxidizes vitamin E.

One experiment published in *Experientia* indicated that feeding weanling mice massive amounts of inorganic sodium selenite produced cataracts. The amount of inorganic sodium selenite used was 3,158 micrograms per kilogram of weight, which is equivalent to 221,000 micrograms for a human. This massive dose compares poorly to the recommended dietary range of 50 to 200 micrograms.

Other researchers have noted similar effects with inorganic sodium selenite injected subconjunctivally (Alagan 1957, Ostadalova 1978). This has been confirmed by a research group at the Mount Sinai School of Medicine of the City University of New York.[9]

The point is obvious. Do not take large amounts of inorganic sodium selenite. Obtain your optimal amounts of dietary selenium from foods, selenium-yeast, seleno-proteins and other organic forms of selenium.

SUMMARY

Selenium deficiency may hasten cataract formation.

☐ Cataracts can be produced in laboratory animals by prolonged selenium deficiency.

☐ The selenium content of cataractous human eye lens is about a sixth that of controls.

☐ Cataracts form by oxidation, and selenium is an effective antioxidant.

☐ Selenium will be of little value by itself—total nutrition is required.

REFERENCES

1 Garner, M. and Spector, A. March 1980. *Proceedings of the National Academy of Science* 77(3).

2 Barber, G. 1973. *Experimental Eye Research* 16:85.

3 Flohé, L. 1974. *Klinisches Wissenschaft* 49:669.

4 Sprinker, L. et al. 1971. *Nutritional Reports International* 4:335.

5 Lawrence, R. et al. 1974. *Experimental Eye Research* 18:563-69.

6 Atkinson, D. 1952. *Eye, Ear, Nose and Throat Monthly* 31(2):79.

7 Heffley, J. and Williams, R. *Proceedings of the National Academy of Science* 71:4164.

8 Williams, R. December 1976. *Executive Health* 13(3).

9 Bhuyan, K. et al. May 1980. Second International Symposium on Selenium in Biology and Medicine, Texas Tech University, Lubbock, Texas.

ELEVEN

Sexual function

T HE greatest aphrodisiacs are good health and abundant energy. Optimal nutrition is a significant factor in both. This chapter is not on virility, however, but on fertility, sterility and the well-being of the reproductive system.

Incidently, selenium may be a critical factor in your body energy level. In his study of dietary liver necrosis, Dr. Schwarz noted that one of the first signs of selenium deficiency was an impairment of mitochondrial function. Mitochondria are components of human cells that are responsible for the synthesis of energy. Protecting these mitochondria from free radicals is an important function of glutathione peroxidase. There are several respects in which this mitochondrial function may be impaired by free radical damage: the mitochondrial inner membrane, which helps to generate energy, contains large amounts of fats whose structure is of key importance to respiratory-energy function. If these fats are peroxidized by free radical attack, impaired function almost certainly results. (It is no accident that the inner mitochondrial membrane also contains a high concentration of vitamin E.)

Coenzyme Q also plays an essential role in human energy production. It is easily damaged by peroxides and levels of Coenzyme Q are known to decline in selenium deficiency. (Incidentally, Coenzyme Q has a strong immuno-stimulant effect, which may help explain the immuno-stimulant activity of selenium.) Finally, certain important protein compounds of the respiratory-energy chain are lost in selenium deficiency.

The mitochondrial respiratory-energy chain is vulnerable to free-radical attack in a number of ways. This finding takes on special importance in light of studies demonstrating that the energy chain itself can produce free radicals. Optimal selenium nutrition improves mitochondrial efficiency, thus maximizes human energy levels.

In recent years physicians have reported alarming increases in both male impotence and sterility, even among young men. In the absence of psychological or organic problems, poor nutritional status could be the sole cause. When one is ill or fatigued as a result of malnutrition, the sex drive naturally diminishes. It is difficult to be amorous while sound asleep in front of the television set.

The fact that malnutrition has an effect on human reproductive capability is suggested by several research findings.[1] The interrelationships and separate effects of selenium and vitamin E on the reproductive process have been the subject of continuing research. Vitamin E deficiency in rats during pregnancy, it has been found, leads to fetal resorption that cannot be prevented by selenium.[2,3,4]

Dr. W. Hartley of New Zealand reported in 1960

that selenium supplementation in sheep significantly improved reproductive efficiency.[5] And Dr. S. Gunn determined in 1967 that selenium is a potent protector against testicular injuries caused by cadmium.[6]

Dr. U. M. Cowgill of the University of Pittsburgh also considered the epidemiological link between selenium and fertility. "An hypothesis was proposed that the birth rates in the high Se [Selenium] regions of the Continental United States were greater than those found in the low Se regions. High regions were separated from medium and low Se areas on the basis of that element's concentration in forage crops, specifically alfalfa. This plant takes up Se in relation to the Se concentration in the soil.

"The purpose of this paper is to test this hypothesis critically. In order to do so natality data were gathered from publications of the National Center for Health Statistics and population data were used from publications of the U.S. Census. Data show live births per 1000 live population, birth rate per 1000 women 15-44 years of age, birth rate subdivided 5 and 10 year age groups per 1000 women and further subdivided racially and still further subdivided into population groups such as urban and rural and metropolitan and nonmetropolitan counties. All of these data are shown for the low, medium and high Se regions. In most cases the pattern that emerges is consistent, namely that the high Se regions exhibit a higher birth rate than the medium and low Se areas."[7]

Infant survival rates also vary in agreement with the hypothesis that low-selenium areas have higher infant mortality.

Without exception, animals that are extremely deficient in selenium are sterile. Such animals produce fewer sperm cells and the majority of the few sperm cells that are produced have poor motility (ability to travel to fertilize an egg cell). Often such sperm cells have their flagellum (propelling mechanism) break off completely.

Sperm cells contain relatively high amounts of selenium. Thus significant amounts may be lost in sexual intercourse.

Selenium that has been purposely made radioactive so that it may be traced throughout the body after it has been eaten has shown that 25 to 40 percent of the selenium concentrates in the male sex organs. Selenium also concentrates in the female sex organs, but to a lesser extent.

Drs. S. H. Wu, J. E. Oldfield, P. D. Whanger and P. H. Weswig of Oregon State University have determined that selenium-deficient rats produced non-motile spermatoza. The majority of the sperm cells broke near the tail.[8,9] The researchers concluded that the role of selenium on the production of spermatoza is specific and cannot be substituted either by vitamin E or by other antioxidants tested.

These findings were confirmed and further elucidated by Drs. Harold Calvin and Edith Wallace of the Center for Reproductive Sciences of Columbia University.[10] Drs. Kenneth McConnell and Robert Burton of the University of Louisville School of Medicine have studied the selenoproteins formed in the testes.[11] They note that selenium tagged for tracing by radioactivity accumulates there as a stable protein over a period of several weeks in contrast to several other organs in which selenium levels peak in hours and then fall. This indicates a major role played by selenium in the testes.

SUMMARY

☐   Selenium is essential for reproduction.

☐ Selenium is needed for energy production in the mitochondria and fatigue depresses the sex drive.

☐ Normal sperm cells contain high levels of selenium.

☐ A prominent sign of selenium deficiency in animals is infertility.

☐ Selenium deficiency reduces sperm production and impairs sperm motility.

☐ Birth rates in the U.S. are directly proportional to dietary selenium levels.

REFERENCES

1 Bongaarts, J. May 9, 1980. *Science* 208:564-69.

2 Evans, H. and Bishop, K. 1922. *Science* 56:650.

3 Christensen, F. et al. 1958. *Acta Pharmacological Toxicology* 15:181.

4 Harris, P. et al. 1958. *Proceedings of the Society for Experimental Biological Medicine* 97:686.

5 Hartley, W. et al. 1960. *New Zealand Journal of Agriculture* 101:343.

6 Gunn, S. et al. 1967. In *Selenium in Biomedicine* eds. Muth, Oldfield and Weswig. Greenwich, Connecticut, AVI, p. 395.

7 Cowgill, U. May 1976. *Proceedings of the Symposium on Selenium-Tellurium in the Environment.* Pittsburgh, Pennsylvania: Industrial Health Foundation.

8 Wu, S. et al. 1969. *Proceedings of the American Society for Animal Science* 20:85.

9 Wu, S. et al. 1973. *Biological Reproduction* 8:625-29.

10 Calvin, H. and Wallace, E. May 1980. Second International Symposium on Selenium in Biology and Medicine, Texas Tech University, Lubbock, Texas.

11 McConnell, K. and Burton, R. May 1980. Second International Symposium on Selenium in Biology and Medicine, Texas Tech University, Lubbock, Texas.

TWELVE

Selenium detoxifies pollutants

Our modern world inundates us with heavy metal pollutants. Lead pollution from burning gasoline causes brain damage. Cadmium pollution from cigarettes causes high blood pressure. Mercury pollution from industry and pesticides is extremely toxic. Fortunately certain nutrients, including selenium, help protect us from this toxic onslaught.

Selenium has the ability to detoxify a number of heavy metals via several methods. Selenium has a high affinity for these toxic metals, binding to them and rendering them harmless. In foods selenium binds to pollutants to prevent their absorption. Selenium increases the excretion of arsenic into bile.[1] Selenium alters the tissue disposition of mercury, cadmium, silver and thallium. There is evidence that selenium exerts its protective effect in other, as yet unknown, ways.

Lead

Ancient man had barely a trace of lead in his bones. The body of modern man, however, contains thousands of times more because lead is ubiquitous; the biggest offender is gasoline burning. It causes many disorders from fatigue to mental retardation to death.

Dr. Orville Levander of the Nutrition Institute of the U. S. Department of Agriculture has found that selenium protects against lead poisoning.[2] Dr. Levander noted that the vitamin E status is more important than the selenium status in protecting against lead poisoning, but excessive dietary selenium did protect partially against lead poisoning in vitamin E-deficient rats.

Cadmium

Cadmium from cigarette smoke, industrial waste, tap water, and dozens of other sources is polluting every citizen of industrialized nations. Rats develop high blood pressure when given cadmium, but do not do so when equivalent amounts of selenium are given at the same time, according to researchers at the University of Rochester.

Dr. Henry A. Schroeder of the Trace Mineral Laboratory of Dartmouth Medical School notes that although zinc and some chelating agents are effective in removing cadmium from the body, selenium is 100 times as effective as zinc.

Mercury

In the early 1970s, some kinds of fish were occasionally banned by the FDA because they contained dangerous levels of mercury. In 1972 Dr. W. G. Hoekstra and his colleagues at the University of Wisconsin found that tuna containing mercury was less dangerous to man than previously suspected. They found that selenium, which also concentrates in tuna, reduces the toxicity of ingested mercury.[3] In one study, for example, they found that rats given water containing 10 parts per million mercury died within six weeks, while similar rats that were also given 0.5 part per million of selenium survived. The researchers suggested that selenium and mercury combine in the blood to reduce the biological availability of both.

Later, Dr. P. Whanger's group at Oregon State University found that selenium injected into mercury-poisoned rats was protective through another mechanism. Injected selenium prevented the mercury from binding in its normal fashion with low-molecular weight proteins by diverting it to less critical components.[4]

SUMMARY

Several nutrients, including selenium, help protect us against heavy metal poisoning. We are all subjected to these pollutants, so we must be sure to get optimal amounts of selenium, sulfur-containing amino acids such as cystine and methionine, vitamin E, vitamin C and zinc. By optimal amounts we mean more than that which is needed just for nourishment, because some will be bound to the pollutant and unavailable to the body.

REFERENCES

1 Whanger, P. 1976. *Proceedings of the Symposium on Selenium-Tellurium in the Environment*. Pittsburgh, Pennsylvania: Industrial Health Foundation, pp. 234-52.

2 Levander, O, Morris, V. and Ferretti, R. 1977. *Journal of Nutrition* 107(3):378-82.

3 Ganther, H. et al. 1972. *Science* 175:1122.

4 Chen, R. et al. 1974. *Pharmacological Research Commentary* 6(6):571-9.

THIRTEEN

The
experts
speak out

As much of research on the apparent role of selenium in preventing the major killer diseases comes from the laboratories of Dr. Gerhard Schrauzer and Dr. Raymond Shamberger, I have asked each of them to write a brief overview of their research and thoughts on the role of selenium.

I have also included similar remarks by Dr. Klaus Schwarz, to whom this book is dedicated, and a review of the essentiality of selenium in human health by Dr. Doug Frost.

Let's begin with Dr. Gerhard Schrauzer, Professor of Chemistry, University of California at San Diego.

Dr. Schrauzer:

My interests in selenium as an anticarcinogenic agent were triggered by a study of the catalysis of electron transfer reactions involving organic dyes and thiols (sulfur compounds) as reducing agents, which began in 1966. In 1969, we noticed a paper by R. J. Savignac, J. C. Gant and I. W. Sizer[1], quoted by R. J. Winzler[2], as well as another paper by M. M. Black and F. D. Speer[3], dealing with dye-decolorization tests with human plasma. These tests were recommended for cancer diagnosis, since the plasma from cancer patients differed considerably in its ability to decolorize methylene blue, for example, from normal plasma. According to Black, the "methylene blue reduction test" was sufficiently accurate for cancer diagnosis. However, later studies by N. Eriksen et al[4] indicated that the test was insufficiently specific, and hence it was forgotten, until we worked on it again in 1966-1970.

We studied this test and found that it was actually a crude assay for plasma *selenium*. From this we concluded that selenium could have anticarcinogenic effects. I always wanted to find a safe, effective method of cancer prevention and was immediately drawn to the idea of "nutritional prophylaxis." It was necessary to find a suitable animal tumor model system for our subsequent experiments. We selected the C_3H/St mice which carry the Bittner Particle because we felt that it represented a good model of human breast cancer. In the meantime, this was confirmed through studies of Spiegelman and coworkers.[5] These authors found the same antigenic components in organs and tissues of human breast cancer patients as are present in Bittner-virus infected C_3H mice. Bittner-virus-induced tumors are some of the hardest to prevent, incidentally, but we succeeded with selenium at subtoxic concentrations and completely natural maintenance conditions. All other workers either investigated chemically induced or transplanted tumors. Such tumors are not necessarily models of human tumors, and since the carcinogens have to be applied often or at toxic levels, the relationship to the human cancer mortality experience is tenuous because one deals with poisoned animals which are also sick otherwise. Bittner-virus-bearing animals are not sick; they are indistinguishable from normal mice.

We are, of course, also interested in cancer epidemiology and have published in this field, as you surely know. The results of this work also support the cancer-protecting effect of selenium. Our work on selenium started independently from, and simultaneously with, that of Shamberger and his colleagues.

Our present research still concentrates on selenium as an anticarcinogenic agent. However, we also are investigating the effects of selenium-antagonistic elements (arsenic, lead, cadmium, zinc, copper, for example). If selenium is to be used for cancer prevention, it is important that supplements are used which are acceptable. We therefore study the various ways of selenium supplementation and have convinced ourselves that selenium must be present in the form or forms normally found in foods. The proper amount of selenium is difficult to get unless one really lives on a well-balanced diet in a "selenium-adequate" region. For prophylactic purposes a total selenium intake of 300 micrograms per day per adult would seem adequate, provided that the selenium is ingested in bioavailable form and the intake of selenium-antagonistic elements or compounds is not too high. I think that the provisional RDA of 50-200 micrograms is too low. It is based too much on the philosophy that the current average U.S. nutritional selenium intakes are "adequate." A better range would be 300-400 micrograms. One should also not just call it "selenium" but should instead make sure that it is to be "biologically active, natural and organic." (The presently available selenium yeast supplements meet these criteria.)

People should get the proper amount of selenium as a unique way to ward off environmental stress, which includes certain toxic heavy metals, mutagens, cancer-causing substances, radiation, peroxides, ozone, alkylating agents, etc. By lowering the number of mutations due to environmental stress, the natural resistance of the organism is increased not only against cancer but also against other degenerative diseases (congestive conditions of the heart, muscular degeneration, arthritic conditions, cataract formation, etc.).

Preferred dietary sources are: cereals, organ meats (liver, kidney), seafoods (shellfish contains sometimes not only a lot of selenium, but also of arsenic). "Cereals" are whole wheat grains, preferably, and whole wheat cereal products. 100 grams of a good whole wheat grain bread may furnish as much as 60 micrograms of selenium. But bread is unfortunately rich in calories, so supplementation is a better way to achieve adequate selenium intakes if one also wants to stay slim.

150-200 micrograms of selenium per day is a good maintenance dose for all ages. (I personally take 200 micrograms per day, occasionally 300-400 micrograms.)

There are no toxicity problems at a regular dosage of up to 800 micrograms of (selenium-yeast-) selenium per day even over extended periods. Even 2,000 micrograms can be taken for some time without harm, but this is about the limit.

Dr. Shamberger:

I first became interested in selenium in about 1964 because I observed that 7, 12-diemthylbenzanthracene-croton oil induced skin tumors regress after about sixteen weeks. This meant that something activated the lysosomal enzymes. Because the lysosomal membranes were thought to be stabilized by the antioxidants vitamin E and selenium, these two as well as others were tested in the DMBA—croton oil system. Soon it became apparent that selenium had special protective properties when applied to the mouse skin tumor system. In about 1968 I decided to see if that effect were nutritional and found that dietary selenium also inhibited the DMBA-croton oil skin tumor testing system.

Our present research interests are to see if selenium might also inhibit or reduce the incidence of human cancer and heart disease. Selenium is also known to prevent heart muscle degeneration in animals. Perhaps a proper amount of dietary selenium will prevent human heart disease and cancer. However, no definite disease in humans has been related to selenium deficiency.

I agree with the provisional RDA of 50-200 micrograms per day even though this does not consider individual absorption rates and bioavailability. Selenium has interrelationships with other trace nutrients such as cadmium, mercury, lead and zinc. These metals may compete for the same binding sites.

My preferred selenium source is wheat bread and wheat products. Seafood is another good source of selenium. Because we already take in about 100 micrograms per day, I don't recommend taking more than another 100 micrograms as supplements. In general, the organic form is considered less toxic than the inorganic form but at 100 micrograms per day no real difference should be seen. Either the organic or inorganic form would be all right. Most beneficial effects on animal disease have been observed with the inorganic form, but most supplements are in the organic form.

The amounts of selenium to bring about human toxicity are not known. Several adults have taken up to 1000 micrograms in the organic form for several months and had no ill effects. Toxicity can occur and has been observed in animals. Supplements should be taken as directed because of possible toxicity in very large doses.

Selenium at first was thought to be a carcinogen among scientists due to some early experiments with faulty design. These results were obtained at very large doses which were in the toxic range. Since then about 20 experiments with different carcinogenic testing systems have been done. They show that selenium has a dietary effect against cancer. The organs protected were mainly the breast, colon, skin and liver. Misconception among the lay people could arise from the same early misconceptions that scientists had. Selenium is thought to work by being a cofactor for the enzyme glutathione peroxidase which removes cellular peroxides which can cause cellular damage by peroxidation.

The future research outlook for selenium is very promising in regard to preventing human cancer and heart disease. One approach might be a large scale government trial done over a ten-year period in a higher risk group using selenium supplementation. Until this is done investigators could look at high and low blood seleniums in different populations and compare their disease incidence and mortality.

I dedicated this book to Dr. Klaus Schwarz for many reasons. Two of the main reasons are: First, his many important contributions to our knowledge of nutrition and health, and second, his many contributions to my own research. His wife, Joyce, has written the foreword to this book.

In the late 1960s, I followed Dr. Schwarz's leads on the selenium-containing "Factor-3." In the late 1970s, I was still following the lead of Dr. Schwarz and his colleague Dr. Walter Mertz (now head of the Human Research Section at the U. S. Department of Agriculture) on chromium and the glucose tolerance factor (GTF). At this writing, I am trying to isolate "Factor G," the growth factor detected by Dr. Schwarz.

The highlights of Dr. Schwarz's pioneering career were published in the National Institutes of Health *Record*.

"Dr. Klaus Schwarz, renowned investigator in trace element nutrition and a former scientist with the National Institute of Arthritis, Metabolism, and Digestive Diseases, died January 30, 1978 in Los Angeles.

"At the time of his death, Dr. Schwarz was chief of the Laboratory of Experimental Metabolic Diseases, V.A. Hospital, Long Beach, Calif., and associate professor, department of biological chemistry, UCLA School of Medicine.

"Dr. Schwarz, who came to the National Institutes of Health from Mainz, Germany in 1949 as a research fellow, eventually became chief of the Liver Disease Unit, Laboratory of Nutrition and Endocrinology, NIAMDD.

"In this capacity, he discovered a water-soluble factor, "Factor 3" that prevented liver necrosis in rats and subsequently identified the factor as selenium. He was assisted in this work by Dr. Calvin Foltz, now with the Laboratory of Chemistry, NIAMDD.

"Dr. Schwarz's work on selenium led to the recognition that this element was deficient in soils from many areas of the world and that large economic losses of livestock and poultry could be prevented by dietary supplements of selenium.

"As an outgrowth of the isolation of selenium, a second essential trace element, chromium was discovered in Dr. Schwarz's NIAMDD laboratory in conjunction with Dr. Walter Mertz, now at USDA.

"In 1963 Dr. Schwarz moved to the V.A. Hospital in Long Beach and expanded his trace element research there and at UCLA as a grantee of NIAMDD. He showed that tin was also an essential trace element for rats, and that growth promoting effects could be demonstrated for vanadium."

It would be a great task to describe all of the contributions of Dr. Schwarz. I have abstracted some of Dr. Schwarz's own words from comments he once made at the conclusion of the Selenium Symposium held at the University of Notre Dame.[6]

Dr. Schwarz proved that selenium was a protective factor. I hope you can feel the excitement of nutritional research in his words.

Dr. Schwarz:

In my scientific career the selenium story goes back to 1939, when, as a young man in Richard Kuhn's laboratory at the Kaiser Wilhelm Institute (now Max Plank Institute) in Heidelberg, I detected dietary liver necrosis as a new deficiency disease in rats. With only few interruptions, work on dietary liver necrosis has been continuously in progress in our laboratory since then. As we now know,

this disease is of complex origin, like most, but not all
selenium or vitamin E responsive disorders. It is induced
by the simultaneous lack of selenium and vitamin E from
the diet. Relatively early it became evident that the sulfur
amino acids also exerted some protective effect. In retrospect, therefore, the diverse types of diets used in studies
on dietary liver necrosis were rather low in sulfur amino
acids and deficient in tocopherol and selenium.

Discovery of dietary liver necrosis

I discovered liver necrosis when I wanted to show that
the so-called Factor H' was necessary for the rat. The late
1930s were the heydays of vitamin research. Following
the clarification of the structure of riboflavin, a close race
developed between our laboratory in Heidelberg and several groups in the U.S. in the isolation and clarification of
the chemistry of vitamin B-6. This was followed by attempts to isolate the "filtrate factor" which essentially was
identified with and eventually identified as pantothenic
acid. Similar to groups in America, we had found that not
only animal assays, but also bacterial growth systems could
be used to test for various B vitamins. One of the first
tasks which I tackled was the isolation of the filtrate
factor, using the animal assay and the bacterial growth
tests side by side. In highly purified pantothenic acid
concentrates I discovered a novel growth factor for lactic
acid bacteria which we designated Factor H'. It could be
separated from pantothenic acid by means of phospho-
tungstic acid. In an attempt to show that Factor H' was
required by animals, I made highly purified casein diets.
Eighty percent of the first group of weanling rats main-
tained on the casein VI diet died within 32 days from

sudden, fatal disease which turned out to be liver necrosis. It immediately became evident that this disease was a new dietary deficiency disease.

Already in our first animal test with casein VI evidence was obtained that this disorder was not related to Factor H'. However, protective effects against liver necrosis were elicited by a variety of nutrients, such as wheat germ, wheat bran, whey and brewer's yeast. Factor H' was isolated and identified as p-aminobenzoic acid. This development, little noticed in this country, was independent of the findings by Woods and Fildes who described p-aminobenzoic acid as a naturally occurring metabolite antagonizing the sulfanilamides.

In contrast to most other workers in the field at that time, I supplemented my diets from the beginning with choline and all other vitamins then available. Thus, I was fortunate enough to deal with a system which produced liver necrosis, per se, without any other associated liver lesions, while in other studies, e.g., those by Paul Gyorgy here in this country, and by Himsworth and collaborators in Britain, complications arose. Their diets produced pathological changes where liver necrosis, on one hand, and fatty infiltration and cirrhosis were superimposed. This confused the issue for quite a while since the two diseases have completely antagonistic dietary and metabolic relationships. A definitive clarification of these problems was reached at the Symposium on "Nutritional Factors and Liver Diseases" which I organized for the N.Y. Academy of Sciences in 1953. This meeting, for the first time, brought together the various groups working on experimental liver diseases of nutritional origin, and those researchers who worked in tropical and subtropical countries on diseases, such as Kwashiorkor, together. It became quite clear that factors which protect against liver necrosis, at least in experimental models, often enhance the development of liver cirrhosis, and vice versa.

Liver necrosis on deficient diets had been observed by many others, especially in rats on low protein rations. The earliest publication is probably that of T. E. Weichselbaum. He fed low cystine diets, observed "hemorrhages throughout the liver," and found that L-cystine on liver necrosis was then more clearly established by F. S. Daft, W. H. Sebrell and R. D. Lillie. In our early experiments cystine had very little, if any effect. Studies with sulfur amino acid supplements very often produced contradictory results. Himsworth and collaborators in Britain, for instance, first identified natural L-methionine as the active factor, then found synthetic methionine to be ineffective while natural L-cystine was active. These discrepancies became explainable after the discovery of selenium. One atom of selenium in form of Factor 3 affords as much protection as 700 to 1000 molecules of vitamin E in preventing dietary liver necrosis in the rat, multiple necrotic degeneration in the mouse, and exudative diathesis in the chick. L-cystine, in turn, possesses only a very small fraction of the potency of tocopherol; to elicit 50 percent protection, about 0.4 percent must be added to the diet. The biopotency of Factor 3 is so high that a trace contamination of one atom of Factor 3-selenium among 355,000 atoms of sulfur could account for the biological potency of the sulfur amino acid. This amount is indeed found in commercial L-cystine, which is manufactured from horn, hoofs and hair and contains up to 2 ug of selenium per g. Procedures which remove the selenium from the cystine also eliminate the biopotency.

Methionine and cystine, when clearly free from traces of selenium, do not prevent liver necrosis. However, they delay its onset when supplied at levels exceeding the requirement for growth. The delay is the result of a sparing effect on the requirement for vitamin E. The sulfur amino acids reduce the vitamin E requirement to only one tenth of the level normally necessary for protection. Thus they exert a modulating effect on the condition. In essence, however, dietary liver necrosis is the result of the simultaneous lack of Factor 3-selenium and vitamin E.

Having found in 1940 that in wheat germ, bran, whey and especially brewer's yeast, a factor or factors were present which prevented dietary liver necrosis, we embarked on experiments to identify the active compounds, using prevention of dietary liver necrosis as an assay. Especially wheat germ contained a great deal of protective activity; we, therefore, used it as a starting material. Within one year a fractionation scheme had been developed. The active substance was found to be lipid soluble, unsaponifiable, not a steroid, and purifiable by chromotography on Al_2O_3 columns. Practically re-isolating vitamin E, we finally ended up with concentrates consisting to 50–65 percent of a-tocopherol before we realized that the protective agent from wheat germ was indeed vitamin E.

We had not failed to put tocopherol into the diets. However, we had miscalculated the amount of which we thought to be needed by dividing the dose protecting against absorption sterility by the number of days of pregnancy in the rat. Thus, our diets actually provided only 5 or 10 percent of the vitamin E level required for protection against the liver disease.

At any rate, this work established vitamin E as a protective factor against dietary liver necrosis in 1942. Because of the high incidence of epidemic hepatitis in the army, the German high command took a special interest in this finding and delayed publication until 1944. Because of the international blackout prevailing at that time this discovery remained relatively unknown in this country until after the war. The development led to my being in this country since the New York Academy of Sciences organized a symposium on vitamin E in 1949 under Karl E. Mason. After presenting these results at the symposium I was invited to join the National Institutes of Health.

That a third factor, other than L-cystine and vitamin E, was involved in the prevention of dietary liver necrosis became apparent very early in our studies: Already in 1943 it was clear, for instance that in preparation of casein VI from vitamin free Hammersten casein, a hitherto unidentified factor had been eliminated which protected independently from cystine and vitamin E. Conclusive proof for the factor, designed Factor 3, was obtained only after 1949. In the final proof for the existence of Factor 3 a peculiar difference between brewer's yeast and torula yeast, grown commercially on sulfite liquor or similar media, has played a decisive role. In Europe, it was found during World War II that torula yeasts could be used as sole source of protein in diets which induced necrotic liver degeneration, but in the U.S. workers were unable to duplicate these findings with American yeasts. Some laboratories resorted to imported British baker's yeast to produce the disease, until an American grown torula yeast was found to be useful for the purpose. Diets containing 30 or 50 percent torula yeast as the sole source of protein have been in use since then for experiments with rats, chicks and other species. Convincing proof for the existence of Factor 3 was obtained by supplementation of a few percent of brewer's yeast to this ration. The addition afforded complete protection. The same effect was elicited by small supplements of "vitamin-free" caseins, shown to contain Factor 3. Comprehensive

studies were carried out in the following years to determine the natural distribution of Factor 3, to devise suitable fractionation and isolation procedures for it, and to
clarify its role in intermediary metabolism. Factor 3 was
found to be widely but unevenly distributed in nature.
Even in kidney, a potent natural source, it was present
only in minute absolute amounts. Like many other micronutrients, Factor 3 was bound to its natural environment, apparently to protein, and could not be extracted
except after autolysis or hydrolysis.

At first, Factor 3 appeared to be a new vitamin. The
discovery of organically bound selenium as the essential
constituent of Factor 3 was made in our laboratory in
1957, following seven years of intensive attempts to isolate
this elusive agent. Indeed, little did I realize what I was
getting into when I started this work. The substance was
clearly organic in nature. We produced over 4,000 fractions and tested them all in groups of ten to twenty
animals against liver necrosis for thirty- to forty-day
intervals, not realizing that the organic compound (or
compounds) which we were dealing with contained a trace
element in bound form. Two years before we identified
selenium as a germane part of Factor 3 it was strongly
suggested by those in charge at N.I.H. that the project be
abandoned. At that time we had already very highly effective preparations but the yields were quite small. I owe
it especially to the insight of Dewitt Stetten that I was able
to bring this work to a definitive conclusion. Eventually
we landed up with 1.3 mg of highly purified material
after using over a ton of pork kidneys in the development
of a complex fractionation scheme. With this 1.3 mg I
experimented in an all-out effort to break the Factor 3
story from January until May 17, 1957 when I finally
identified selenium in my products and correlated the

selenium contents with biopotency. Actually, a peculiar garlic-like odor which was developed by some of my preparations under strongly alkaline conditions led to this discovery.

In retrospect, the best Factor 3 preparations obtained were approximately 10,000-fold as active as the starting material. Quantitative analyses showed approximately 7 percent of the element. If one assumes the molecular weight of Factor 3 to be around 300, these fractions would be roughly 25–35 percent pure. Efforts to isolate the active compound from kidney powder and other natural sources were halted because of the very low yields and the extreme liability of the purified agent. The situation was similar to that encountered today with the active site of glutathione peroxidase.

The only other organic selenium compound of natural origin which existed at that time was selenocystathioneine. It had been isolated by Horn from seleniferous wheat. He worked at the Beltsville Laboratories of the Department of Agriculture and generously supplied me with what amounted to the world's supply of selenocystathioneine. It was one of the first selenium compounds found to be effective.

Already prior to the discovery of selenium as the active principle in Factor 3 preparations it had become evident in our laboratory that this dietary agent was an essential nutrient in its own right regardless of its interrelationship with vitamin E. In rats provided with tocopherol, Factor 3-deficiency causes a distinct disease entity comprising lack of growth, muscular wasting, adrenal atrophy and pancreatic dystrophy. In the chick, Factor 3-active selenium showed a growth effect independent of vitamin E. That Factor 3/selenium did not simply "replace" vitamin E became strongly evident in metabolic studies on respiratory decline of livers from animals during the latent, prenecrotic phase of dietary liver necrosis. Vitamin E showed entirely different effects from Factor 3/selenium in this system. Studies with liver slices from rats on the necrosis-producing diet showed that a metabolic defect, respiratory decline, is present one to two weeks before actual necrosis occurs.

Such liver slices cannot maintain normal oxygen consumption. Addition of vitamin E to the diet, or the intravenous injection of the vitamin ten to thirty minutes prior to extirpation of the liver will completely prevent the breakdown. In liver slices Factor 3/selenium affects respiratory decline to a much lesser degree. In liver homogenates respiratory decline is completely preventable by the addition of near-physiological amounts of a-tocopherol while selenite or selenium compounds show no effect whatsoever on the respiratory failure.

These and other lines of evidence showed clearly that selenium was essential, as such, for warm blooded animals. Nevertheless, attempts were made, by others later on to "prove" the essentiality of selenium, for example, by virtue of its specific effect on pancreatic dystrophy in chicks.

Our findings on dietary liver necrosis were the first to establish the interesting interrelationship between selenium and tocopherol, and to a wider extent the sulfur amino acids. The vast pathological changes typical of selenium and tocopherol responsive diseases can be explained as resulting from a relationship first published in 1957. It postulates that the two agents do not substitute for each other, but that they are essential for alternative pathways of intermediary metabolism. This would explain why the absence of both is necessary for the development of the vast pathological changes which develop in so many different species in a large variety of organs. Through the recent exciting discovery of selenium as the active center of glutathione peroxidase, a part of this hypothesis has become scientific reality. It is still unknown what specific catalytic function is carried out by tocopherol. It is my opinion that it does not merely act as an antioxidant but as a specific catalyst in intermediary metabolism. The identification of selenium as the active site of glutathione peroxidase could also explain why sulfur amino acids exert some protective effect on selenium responsive diseases. Lack of glutathione in livers of animals on diets producing liver necrosis seems to contribute to the seriousness of the metabolic impairment.

Work on selenium in our laboratory has been concerned with several main aspects. We tried to identify the site of action of selenium in intermediary metabolism and around 1970 actually had narrowed our project down to enzymes involving glutathione metabolism; various reasons indicated this substance to be related to respiratory decline. Indeed, a visiting scientist in my group spent almost a year on studies with glutathione dehydrogenase systems. I must blame myself for overlooking glutathione peroxidase. We were at the right spot, so to speak, but we were "barking up the wrong tree." Thus it remained to the splendid work by Rotruck, Hoekstra and others at Wisconsin, and Flohé and collaborators at Tübingen to establish glutathione peroxidase as an enzyme containing selenium as the active site.

The main project on selenium carried out in our laboratory was concerned with the clarification of the relationships between structure and biologically potency of organic selenium derivatives. This large scale effort was made between 1958 and 1973, mostly in close collaboration with Professor Arne Fredga and his associates at the Institute of Organic Chemistry at the University of Uppsala, Sweden. The basic goals of these studies were: (a) To delineate as well as possible the structural properties required for Factor 3 activity, (b) to synthesize effective selenium compounds of low toxicity that could be useful in the treatment of diseases.

The activity of over 850 substances has been determined in the course of this project. Approximately half of them originated in Professor Fredga's Institute; the rest were prepared in our laboratory in Long Beach. We measured the protective effect of each compound in series of experiments, using a scale of different dose levels; prevention of death from dietary liver necrosis in the rat over a forty-day period of experimentation served as the criterion of potency. Selenite-selenium was used as a standard of comparison. Beautiful and interesting regularities were found over a wide range of selenium compounds in the relationship between chemical structure and biological activity.

Following the determination of the structural properties which account for activity in the prevention of liver necrosis we tested most of our compounds for acute intraperitoneal toxicity in the rat and investigated the physicochemical parameters which determine toxicity. Then we prepared selenium compounds which combine in their structure properties which are essential for biological activity with those characteristic for lack of toxicity.

The end result was the development of selenium compounds which are biologically effective but show only a minimum of toxicity. The substances are relatively simple amino acid derivatives of monoseleno diacetic acid. They are not only less than one hundredth as toxic as most other selenium derivatives when applied by injection, they are equally well tolerated when given by mouth or in the diet over long periods of time. Preliminary tests, for instance, have been carried out with selenodiacetyl alanine added to the diet in amounts supplying 25 to 100 mg selenium per 100 g of diet, respectively. Under these conditions rats grew from 50 to over 200 g average weights within fifty days of experimentation. Their livers were normal. Controls receiving 1 and 4 mg percent of selenite selenium in the diet were dead within three to seven days.

It is hoped that these selenium compounds of minimum toxicity may open up new possibilities for the application of selenium in prophylaxis and therapy of human diseases.

Selenium deficiency has been a major economic handicap to animal breeders in many low-selenium regions of the world. Less than twenty years ago, the New Zealand sheep industry was on the verge of financial ruin due to White Muscle disease, the ailment which cripples young animals, often killing them within ten days of birth. White muscle disease was also occurring with increasing frequency among livestock in other areas of the world. Dr. John A. Schmitz, of Oregon State University, estimates losses in the northwestern United States in the 1950s topped $11 million a year.

In some regions, pigs suffered from liver damage and "mulberry heart," a cardiac degeneration culminating frequently in sudden-death heart attacks. Chickens were dying from a fibrous degeneration of the pancreas and from "*exudative diathesis*," a fatal disease resulting from increased capillary permeability. Horses, dogs, and bulls were frequently crippled from unknown causes.

Soon after the discovery of selenium nutritional essentiality, it was discovered that each of these ailments resulted from selenium deficiency. Today, these diseases are effectively prevented or treated by simply adding selenium to feed grains.

The range of symptoms produced in animals by selenium deficiency is amazingly diverse. These include liver damage, pancreatic damage, growth retardation, a form of muscular dystrophy, heart failure, hemorrhage, sperm loss with infertility, hair loss and cataracts. It is apparent that selenium is of the most vital importance to the health of almost all body tissues.

Dr. Schwarz has shown selenium to be an essential nutrient. In 1973 Dr. W. Hoekstra's group at the University of Wisconsin determined that selenium's biochemical role is as a component of the enzyme glutathione peroxidase.[7]

At the 1980 Selenium Symposium, Dr. Doug Frost of Schenectady, New York, gave an overview of the role of selenium in human health. Dr. Frost became interested in selenium while a researcher at the Trace Mineral Laboratory of the Dartmouth Medical School. He has been a major force in selenium research due to his vast knowledge of the subject and his ability to bring this information together in perspective.

Here is an abstract of Dr. Frost's presentation.

Dr. Frost:

The absolute need for selenium as a nutrient was shown for many animal species. With vitamin E, selenium serves as the active part of one form of glutathione peroxidase to minimize aberrant oxidations and free radical damage. With E, or other antioxidants, selenium binds and helps detoxify or sequester various heavy metals.

Because selenium was wrongly impugned as a carcinogen from 1943 until recently, its apparent anticancer value was masked and is only now under critical study. Animal research has indicated roles for selenium and E to prevent or delay chronic diseases of aging. This may reflect their stimulation of ubiquinone (CoQ_{10}) biosynthesis. Evidence suggests that adequate selenium and E may delay cataract formation, periodontal disease and cardiovascular disease. Evidence from New Zealand showed that selenium is needed in total parenteral nutrition, from Germany that it is needed in highly purified diets for children with metabolic diseases, and from China that it is needed to prevent Keshan disease, a cardiomyopathy in children on meager home-grown fare. Evidence suggests that the selenium adequacy of crops is decreasing, with mounting selenium deficiencies in animals worldwide. This may be due in part to the increasing SO_2 [sulfur dioxide] emanation and resultant diminution of selenium uptake by plants. Heart, liver and muscle myopathies and sudden death have been described in sheep, cattle, swine, poultry and even wild or zoo animals.

Whereas veterinary products of selenite selenium and vitamin E have been used to treat selenium-responsive diseases for over two decades and feed additive uses of selenite or selenate have been used to prevent such diseases, human medicine lacks similar products. Human tolerance to such products is well established and any hazards from their possible overuse are easily avoided. Selenium yeast products have been widely sold over the counter to provide up to 150 ug of supplementary selenium per day as selenomethionine. Selenite selenium with vitamin E has medical advantages, as evident in veterinary use.

Evidence from Russia has indicated value of selenium and vitamin E to prevent hypercalcification from too much vitamin D. Calcium and selenium metabolism are interdependent, as seen in the deposition of calcium in white muscle disease. Evidence calls for investigation of possible value of selenium plus E versus bursitis and other forms of arthritis and joint disease. Whether selenium improves visual acuity and has value against diabetic retinopathy also deserves clinical study. Opportunities for prevention or curative applications of selenium, combined with E and other essential nutrients, seem compelling and warrant attention in clinical nutrition and medicine.

The unique value of selenium plus E to reduce oxidative damage and damage from heavy metals, excess nitrites, radiation, chlorinated hydrocarbons, nitrosamines and other carcinogens reflects their enhancement of body defense mechanisms. How they and other essential nutrients interact will be shown by research. Freedom to experiment clinically with selenium is needed to enable Americans to help solve this magnificent puzzle. Benefits to health and extension of useful lifespan will follow.

Now that we know that selenium is essential, how much is enough? Let's examine the Recommended Dietary Allowance (RDA) in the following chapter.

REFERENCES

1 Savignac, R. J., Gant, J. C. and Sizer, I. W. 1945. *American Academy of Advanced Science Research Conference on Cancer*, p. 245-52.

2 Winzler, R. J. 1953. *Advances in Cancer Research* 1:519.

3 Black, M. M. and Speer, F. D. 1950. *American Journal of Clinical Pathology* 20:446.

4 Eriksen, N. et al. 1951. *Journal of American Cancer Institute*
   7:705.

5 Mesa-Tejada, R., Keydar, I., Ramanarayanan, M., Ohno,
   T., Fenoglio, C. and Spiegelman, S. 1979. *Annals of Clinical
   and Laboratory Science* 9:202-11.

6 Schwarz, K. 1976. *Proceedings of the Symposium on Selenium-
   Tellurium in the Environment*. Pittsburgh, Pennsylvania:
   Industrial Health Foundation, Inc., pp. 349-69.

7 Rotruck, J. et al. 1973. *Science* 179:588-90.

FOURTEEN

Recommended daily allowances

W E have just seen that Dr. Klaus Schwarz determined selenium essential for life, and Dr. Hoestra's group discovered how selenium functions. But how much do we need for optimal health? Too little is ineffective and too much of anything is detrimental.

The question is difficult to answer because of the "newness" of selenium nutritional research, the complicated character of selenium compounds themselves, and biochemical individuality. Surprisingly, estimates made by selenium experts do not differ by much. Even the committee that establishes the "official" Recommended Dietary Allowances (RDA) has provided us with a reasonable figure.

The Food and Nutrition Board of the National Research Council has undertaken the task of preparing the RDA. In 1974, they noted that selenium was essential, but they could not establish RDA for the mineral.

In 1976, the RDA committee published a status report on selenium which said:

> Available evidence suggests that a well-balanced diet furnishes about 60 to 120 micrograms of selenium daily. Estimates of typical selenium intakes in the United States average about 150 micrograms per day and diet composites analyzed in Canada provide 98 to 220 micrograms per day. . . . although the selenium content of these foods will depend heavily on where the grains were grown. There is also considerable variation in the selenium content of eggs and dairy products. Fruits and vegetables are generally poor sources of selenium, but some exceptions to this rule include garlic, mushrooms and asparagus . . . should selenium supplements eventually be considered desirable for those persons living in low selenium areas, or for those consuming vegetarian diets, a daily supplement of 50 to 100 micrograms could probably be taken safely.

Elsewhere in the status report the Food and Nutrition Board made the following interesting statement:

"Chronic selenium toxicity would be expected in human beings after long term consumption of 2,400 to 3,000 micrograms daily."

The following year at the American Institute of Nutrition in Chicago in April (1977) members of the RDA committee explained a new concept.

Provisional RDAs have been established for a number of trace elements. Dr. Mertz emphasized that, for many, there is a danger from toxicity of excess amounts as well as a danger from dietary deficiency of insufficient amounts. The provisional recommendations are made in terms of a range of values, rather than single figures.

These ranges have been established on the conservative side, with the upper level far removed from toxicity.

The provisional recommendation for adults for selenium was indicated as 50 to 200 micrograms. Dr. Mertz predicted that further research and development will continue in this field. To improve the estimates of requirements, two things are needed: improved methodology and a better understanding of the interactions between minerals and other nutrients.

In 1980, the National Academy of Sciences published the Ninth Revised Edition of the Recommended Dietary Allowances. The discussion of selenium states: ". . . a range of adequate and safe selenium intake can be estimated by extrapolation from many animal experiments and human balance studies. Although this range is somewhat less precise than a recommended dietary allowance, it can serve to warn against marginal intake from imbalanced diets and against over-exposure from selenium-containing vitamin and mineral preparations now available to the public . . . a range of 50 to 200 micrograms per day is suggested as adequate and safe for adults."

It should be noted that Japanese fishermen consuming large amounts of seafoods exceed 500 micrograms of selenium daily.[1]

When selenium supplements became available in health food stores, the Food and Drug Administration tried to ban them, claiming that selenium was not a nutrient but an untested food additive. And this was after the veterinary group at the Food and Drug Administration had itself proposed the supplementation of animal foods with selenium.

According to the inconsistent FDA, selenium supplementation is good for the health of animals because they are selenium deficient—but humans who eat selenium-deficient vegetation and selenium-deficient animals should not enjoy the advantage of selenium supplements.

The manufacturers and distributors of selenium supplements contested the FDA action. After learning of the need for and safety of selenium supplements, the FDA issued a letter of approval in 1978:[2] "Based upon the FDA's present understanding of the current relevant scientific information, the Agency will not recommend legal action based on food additive charges against products containing safe and suitable forms of selenium, provided the total daily intake (from supplementation) when following directions for use, results in the user receiving 200 micrograms (0.2 milligram) or less of the element from the supplement."

Now that the RDA committee has established an allowance for selenium, it may not be too many more years before the FDA incorporates these guidelines into their official "USRDA."

As noted in the preceding chapter, Dr. Raymond Shamberger is in agreement with the RDA, while Dr. Gerhard Schrauzer, based on his studies presented in Chapter Two, feels that the RDA should be slightly higher. Perhaps a good compromise might be to eat a well balanced diet that has 100 to 150 micrograms of selenium in it and then take 100 to 200 micrograms of selenium as a supplement. Remember, moderation is the rule and ten times this level may put you in the beginning of the toxicity range.

Also keep in mind the teachings of Dr. Roger Williams about biochemical individuality. We are each unique and each one of us may have a nutrient need different from the average.

That leaves us with the problem of trying to get 100 micrograms of selenium daily when the foods we eat are deficient in that element. We will take a close look at this problem in the following chapter.

REFERENCES

1 Sakurai, H. and Tsuchiya, K. 1975. *Environmental Physiological Biochemistry* 5:107-18.
2 Roy, C, February 14, 1978. Letter of Regulatory Guidance, Bureau of Foods, FDA.

FIFTEEN

We need more selenium

In Chapter One, I recommended that the vast majority of Americans consider supplementing their diet with selenium. This recommendation applies to most Europeans, Australians and Chinese too.

Most people need selenium supplements because they are not getting optimal levels of usable selenium from their diet. As our activity level decreases, we cut back on our food intake to control our waistline. We are eating fewer calories, but getting fatter. Former estimates for selenium intake were based on diets containing more food than we are eating today. The fact is, dietary estimates based on "typical" diets are unrealistic. People just do not eat as well as nutritionists like to believe. What seems to be a typical diet to a nutritionist is only an occasional diet for the average American.

The food tables are essentially useless. They describe foods obtained from a specific region that may be different in selenium content than yours. These foods also reflect soil conditions that may have had greater selenium availability than the same area does today because of increased acid rain precipitation.

I am concerned about our dietary selenium because it's not just because the dietary selenium level is so low— but because much of the selenium that is in the diet comes in a form that is poorly absorbed or utilized by the body. It's not just what goes into the mouth that's important—it's what gets incorporated into the cells.

The estimated selenium levels for the typical American diet are based on a disproportional quantity of fresh fish. Unfortunately, much of the selenium in fish is bound to mercury and unavailable to nourish the body.

Some people are ingesting considerable amounts of selenium antagonists—heavy metals such as cadmium, mercury and lead—and thus need more selenium.

Dr. Gerhard Schrauzer has pointed out: "A number of heavy metals as well as arsenic are known to act as selenium antagonists *in vivo*. Since normal foods may contain some of these elements in excess of selenium, the selenium requirements for the maintenance of health may be significantly higher than in a diet which is largely free from selenium antagonists."

Therefore, some diets may require more selenium than expected by studies made under ideal diet circumstances.

Another major factor to consider when discussing

the selenium content of foods is that the selenium available in soil is decreasing.

Soil

Figure 1.1 showed the selenium distribution in American soils. Selenium maps for New Zealand, Great Britain, Scandinavia, and China show similar variances, with appreciable low-selenium areas. Unfortunately, the selenium content of soils is decreasing.

Factors in the soil depletion are: (1) removal by crops; (2) modern techniques of fertilizer production have reduced their selenium content; (3) animal excrements of selenium are not recycled because the excreted selenium is in forms not taken up by plants; (4) Most of the selenium from oil and coal burning ends up as elemental selenium which is not taken up by plants; (5) selenium is continually leached from light soils; and (6) acid rain is reducing selenium availability to plants.

Acid rain

As a result of the combustion of tremendous quantities of fossil fuels such as coal and oil, the United States annually discharges approximately 50 million metric tons of sulfur and nitrogen oxides into the atmosphere. Through a series of complex chemical reactions these pollutants can be converted into acids, which may return to earth as components of either rain or snow. This acid precipitation, more commonly known as acid rain, may have severe ecological impacts on widespread areas of the environment. (See Figure 15.1)

Hundreds of lakes in North America and Scandinavia

have become so acidic that they can no longer support fish life. More than ninety lakes in the Adirondack mountains in New York State are fishless because acidic conditions have inhibited reproduction. Recent data indicate that other areas of the United States, such as northern Minnesota and Wisconsin, may be vulnerable to similar adverse impacts.

While many of the aquatic effects of acid precipitation have been well documented, data related to possible terrestrial impacts are just beginning to be developed. Preliminary research indicates that the yield from agricultural crops can be reduced as a result of both the direct effects of acids on foliage, and the indirect effects resulting from the leaching of minerals from soils.

Far more serious than dead fish and dead trees however are *dead people*. Here is how "acid rain" could contribute to your own early demise: As we convert our power plants from oil-burning to coal-burning, acid rain will increase because of the higher sulfur content of coal and the more lenient emissions standards now in effect. The net result will most likely be that less of the selenium in the soil can be taken up by crops. Writing in the *Annual Reviews of Pharmacology*, Dr. Douglas Frost points out how sulfates from acid rain greatly inhibit the uptake of selenium by plants.[1] Result: the amount of selenium in the average American diet is now significantly less than it was in 1920 or 1930. Dr. Frost adds, "Evidence suggests that diminished selenium availability may be worldwide."

Fertilizers probably have no major influence on plant composition in terms of protein, carbohydrate, fat or vitamin content. These nutrients are influenced primarily by the genetic composition of the seed and the maturity of the plant at harvest. Fertilizers do, however, influence the mineral composition of plants.[2]

FIG. 15.1

Acid rain and deposition sensitivity in United States

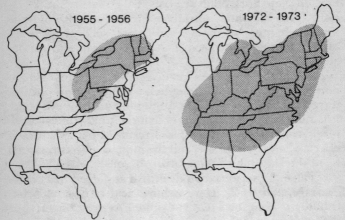

The pair of maps shows the growth of acid rain, while the opposite map indicates those areas of the continental United States that are believed to be sensitive to acid deposition. This map was constructed by examining such factors as chemical composition of soils, climatic patterns and types of vegetation within a given geographical area.

Source: Adapted from Likens, G. E. 1976. *Chemical & Engineering News* (C. V. Cogbill)

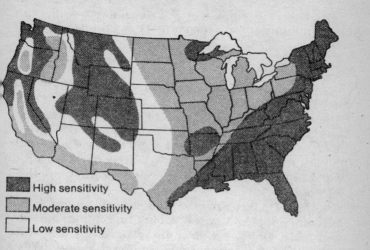

High sensitivity
Moderate sensitivity
Low sensitivity

Remember when the FDA was trying to require that all vitamin and mineral labels bear the ridiculous statement that food nutrients were not affected by soil? At the same time that one branch of the FDA was pushing such non-scientific statements, the FDA veterinary branch was realistic. In an April 27, 1973 news release on selenium, the Food and Drug Administration stated that "levels naturally found in animal feed vary widely depending on the soil in which the feed crops were grown." FDA estimated that about 70 percent of the domestic corn and soybeans do not contain adequate selenium. Such a deficiency can lead to decreased growth, disease, and death of animals feeding on such crops.

In 1967, Dr. J. Kubota determined that over 80 percent of the Corn Belt had soil containing less than 0.1 part per million.[3] In 1969, Dr. Olson confirmed this.[4] The values for selenium ranged from 0.01 part per million in Indiana to 2.03 parts per million in South Dakota. That is a 200-fold variation in the same crop.

Even within a rather small area, the selenium content of corn can be highly variable. Near Gregory, South Dakota, samples varied from 0.14 ppm to 2.03 ppm and near Rushford, Minnesota, from 0.02 to 0.29 ppm. Disregarding the high-selenium states of North and South Dakota, Nebraska and Kansas, 70 percent of the corn had less than 0.05 ppm, while only 3 percent had more than 0.15 ppm.

The FDA veterinary section did win approval for selenium supplements for poultry and swine, but why stop there? Dr. T. J. Cunha of the University of Florida has called for selenium additions to all animal feeds.[5] Unless we fertilize soils or supplement animal feeds with selenium, we will find it impractical, if not next to impossible, to obtain optimal levels of selenium. Until selenium supplementation of all animal feeds is permitted, stockmen should be encouraged to use selenium and vitamin E injections to make their livestock and their customers healthy.

Fertilization is more difficult however. Some crops cannot grow if too much selenium is applied to the soil, while other plants can concentrate selenium to a degree which could be toxic to livestock. Fertilization of alkaline soils with selenium produces oxidation products that are readily soluble and washed away. Fertilization of acid soils with selenium may result in poor selenium uptake.

Availability in foods

Selenium compounds are very volatile. Heat, processing, and cooking diminish the selenium content of foods.[6,7] The refining of grains can destroy as much as 50 to 75 percent of the selenium in the whole grains. Boiling may eliminate about 45 percent of a food's selenium content.

As mentioned, selenium may be bound to metal contaminants in foods thereby preventing assimilation of either. There is also a suggestion that some foods contain selenium antagonists, such as a succinoxidase inhibitor.

Dr. Milton Scott commented, "The very high incidence of selenium deficiency in the eastern part of the United States appears to be due not only to the low selenium content of food, but also to a very low biological availability of selenium in many foods."[8]

Even the estimates that I feel are unrealistically high show that we have a problem. As discussed in Chapter Fifteen, the Recommended Dietary Allowance for selenium is a range of 50 to 200 micrograms. In 1971, the National Academy of Sciences estimated that the average consumption of selenium in the United States is less than 0.1 part per million.[9] This amounts to a range of 20 to 60 micrograms per day. Dr. H. Schroeder estimated the average daily intake of selenium in the United States near the low end of the 60 to 150 microgram range.[10]

Dr. Orville Levander reports estimates for selenium in North American diets ranging from 98 to 224 micrograms a day. (I feel these figures reflect too much fresh seafood which is not consumed in such proportions by the "typical" American. Also selenium is not readily available to the body when it is tied up with mercury.)

Dr. Gerhard Schrauzer estimates an average selenium

intake of 160 to 170 micrograms daily in selenium-rich areas and 100 micrograms in low-selenium areas where most Americans live.[11] He comments: "Dietary selenium intakes in the United States are lower than are those in many other countries ... Dietary selenium intakes of Southern California adults generally range from 90 to 168 micrograms per day. Epidemiological studies suggest that the average dietary selenium intake in the United States is only half that required for optimal protection from malignancies. This protective intake of selenium, 300 micrograms per day, can be obtained from a diet emphasizing whole-grain cereals, seafood, and organ meats only in regions where grains have an adequate selenium content. Dietary supplementation of 150 micrograms per day is suggested for people consuming normal diets in "selenium-adequate" areas. Supplementation of 200 to 250 micrograms per day is suggested for dieters, vegetarians, and those living in "selenium-poor" regions."[11]

The Food and Nutrition Board of the National Research Council has estimated in 1976 that well-balanced diets provide 60 to 120 micrograms of selenium daily. Unfortunately too many people do not eat well-balanced diets for one reason or another, and even less if they live in low-selenium areas.

Researchers at the Ministry of Agriculture in London estimate the average intake of selenium in Britain is approximately 60 micrograms per person per day, of which 50 percent was derived from cereals and cereal products and approximately 40 percent from meat and fish together.[12] The authors also reviewed the following estimates; Canada (110-220), New Zealand (25), Sweden (23-210), Netherlands (110), France (166) and Italy (141).

So much for the average or typical person. You are unique. How much selenium do you eat? The above

FIG. 15.2
Effects of selenium on mammalian life

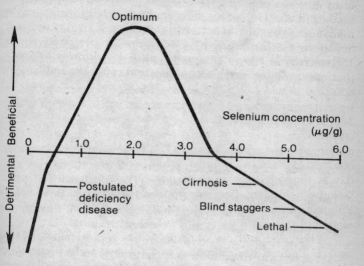

Approximate values expressed as micrograms of total selenium (as selenite) per gram of diet. Toxicity varies with vitamin E and sulfur amino acid levels, as well as the actual selenium compounds present. This graph is for illustrative purposes only, and is not intended to represent safety levels for humans. Selenium toxicity is species dependent. In small animals the hamster is very resistive while the cat is very sensitive.

Source: Passwater, R. 1973. *Fluorescence News* 7(2):11–16.

figures may give you a false sense of security. How much
is optimal for you? See Figure 15.2. *No selenium was found*
in five samples of a *balanced diet* for nineteen-year-old
boys in an FDA survey. The analyses were done by activation analyis which is extremely accurate.[13] Dr. Doug Frost
has noted the scarcity of selenium in prepared meals
generally.[14] I have analyzed typical TV dinners by spec-
trophotofluorometric techniques and could not find any.[1]

Food tables

How much selenium you obtain from your diet obviously
depends upon how much selenium in the foods you eat
and the form of that selenium. The availability of selenium is determined by the food itself, but how much
selenium is present in the food depends upon how much
selenium is in the soil and how available it is.

The following food tables are not representative. They
serve to help you choose selenium-rich foods. Foods generally contain selenium in the following order: organ
meats; fish and shellfish; muscle meats; grains and cereals; dairy products; fruits and vegetables.

The National Academy of Science reports that 5
micrograms of selenium is supplied by 2¾ ounces of
lobster, 4 ounces of cod, 4 ounces of beef liver, and 1½
ounces of beef kidney.

Food grains may store up to about 20 parts per
million from selenium-rich soils. So can onions. But most
vegetables fail to grow after reaching 2 parts per million.
Check your diet by seeing how many selenium-rich foods
you eat. The following tables may help you decide whether
or not you need to supplement your diet with selenium.

TABLE 15.1

Relative food levels of selenium (these values are not representative because they vary from soil to soil)

Selenium content of vegetables and fruits

Product	ug Se/gram
Vegetables	
Carrots	
Fresh	0.022
Canned	0.013
Cabbage, fresh	0.022
Cauliflower, fresh	0.006
Corn	
Fresh	0.004
Canned	0.003
Garlic, fresh	0.276
Green pepper, fresh	0.006
Green beans	
Fresh	0.006
Canned	0.009
Lettuce, fresh	0.009
Mushroom	
Fresh	0.122
Canned	0.109
Onion, white, fresh	0.015
Potatoes	
Sweet, fresh	0.007
White, fresh	0.003
White, canned	0.007
Radish, fresh	0.042

Selenium content of vegetables and fruits

Product	ug Se/gram
Tomatoes	
Fresh	0.005
Canned	0.010
Turnips, fresh	0.006
Mean excluding mushroom and garlic	0.010
Fruits	
Apple, fresh, peeled	0.003
Applesauce, canned	0.002
Banana, fresh, peeled	0.010
Orange, fresh, peeled	0.014
Peach	
Fresh, peeled	0.004
Canned	0.004
Pear	
Fresh, peeled	0.006
Canned	0.002
Pineapple	
Fresh	0.006
Canned	0.008
Mean	0.006

Selenium content of grains and cereal products

Product	ugSe/gram
Barley cereal	0.643
Bread	
White	0.280
Whole wheat	0.676
Corn flakes	0.024
Flour	
White	0.187
Whole wheat	0.627
Noodles, egg	0.662
Oat breakfast cereal, prepared	0.451
Oats, quick	0.114
Rice	
Polished	0.334
Brown	0.383
Rice breakfast cereal, puffed, prepared	0.026
Wheat cereal	0.241
Wheat breakfast cereal, prepared	0.110
Mean excluding corn flakes and rice cereal	0.387

Selenium content of miscellaneous products

Product	ugSe/gram
Egg	
Yolk #1	0.174
Yolk #2	0.192

Selenium content of miscellaneous products

Product	ugSe/gram
White #1	0.057
White #2	0.046
Saccharin	0.005
Sugar	
Brown	0.012
White	0.003,

Selenium content of dairy products

Product	ugSe/gram
Cheese	
American, processed	0.090
Cottage	0.054
Swiss	0.101
Cream, table	0.005
Cream substitute	0.034
Milk	
Evaporated, canned	0.012
Skim	0.045
Skim, powdered, dried #1	0.098
Skim, powdered, dried #2	0.243
Whole homogenized	0.013
Mean	0.069

Selenium content of meats and seafoods

Product	ugSe/gram
Meats	
Beef	
Round steak	0.363
Ground	0.208
Liver	0.454
Kidney	1.41
Pork	
Chop	0.217
Kidney	1.89
Lamb	
Chop	0.172
Kidney	1.38
Chicken	
Breast	0.106
Leg	0.121
Skin	0.154
Mean excluding kidneys	0.224
Seafoods	
Lobster tail	0.634,
Shrimp, shelled, deveined	0.572,
Cod, fillet	0.465,
Flounder, fillet	0.335,
Oysters	0.646,
Mean	0.532

Selenium content of strained baby foods

Product	ugSe/gram
Beef	0.118
Chicken	0.112
Lamb	0.123
Liver	0.247
Pork	0.114
Carrots	0.002
Green beans	0.005
Peaches	0.003
Pears	0.002
Oatmeal cereal with applesauce and bananas	0.026
Rice cereal with applesauce and bananas	0.019
Vanilla custard pudding	0.016,
Mean	0.068

Source: Morris, V. and Levander, O. 1970. *Journal of Nutrition* 100:1383-88.

TABLE 15.2
Selenium content of feedstuffs[a,b]

Feedstuff	Source (if known)	Median value (mg. Se/kg. dry material)
Alfalfa meal	Eastern U.S.	0.10
Alfalfa meal	Germany	0.09
Alfalfa meal	Plains states	0.38

Feedstuff	Source (if known)	Median value (mg. Se/kg. dry material)
Alfalfa hay	Germany	0.06
Bakery by-product	New York City area	0.4
Barley	Eastern U.S.	0.1
Barley	Midwestern U.S.	0.3
Barley	Germany	0.19
Barley, steamed rolled	Wisconsin	0.4
Beans, red kidney	New York State	0.03
Beans	Germany	0.22
Blood meal	Germany	0.07
Bone meal	Germany	0.01
Bread, white[c]	U.S.	0.278
Bread, whole wheat[c]	U.S.	0.665
Brewers dried grains	Midwestern U.S.	0.7
Cassava meal (tapioca meal)	—	0.1
Cheese, American processed[c]	U.S.	0.09
Cheese, cottage[c]	U.S.	0.052
Clay, colloidal	Germany	0.18
Clover, red	Germany	0.11
Corn, yellow dent	Lafayette, Indiana; New York State	0.025
Corn, yellow dent	Midwestern U.S.	0.05
Corn, yellow dent	Nebraska; South Dakota	0.38
Corn, yellow dent	Germany	0.11
Corn, Opaque-2	Lafayette, Indiana	0.038
Corn and cob meal	Midwestern U.S.	0.04
Corn distillers dried solubles	Kentucky	0.5
Corn fermentation solubles, dry	Eastern U.S.	0.91

Feedstuff	Source (if known)	Median value (mg. Se/kg. dry material)
Corn fermentation solubles, liquid	Eastern U.S.	0.09
Corn flakes (cereal)[c]	U.S.	0.026
Corn gluten feed	Eastern U.S.	0.2
Corn gluten meal, 60% protein	Midwestern U.S.	1.15
Cottonseed cake	—	0.06
Crab meal	Eastern U.S.	1.3
Eggs, fresh, yolk[c]	U.S.	0.184
Eggs, fresh, white[c]	U.S.	0.050
Eggs, whole dried	Eastern U.S.	0.35
Eggs, whole dried	Germany	1.0
Fish, carp	Germany	1.37
cod	Germany	1.24
flounder	U.S.	0.24
pike	Germany	1.66
red snapper	U.S.	1.96
salmon	Western U.S.	0.41
trout	U.S.; Germany	1.3
tuna, canned	Japan	0.6
Fish meal, anchovetta	Chile	1.6
anchovetta	Peru	1.29
herring	Maine	1.5
herring	British Columbia	1.8
herring	Norway	2.45
menhaden	Gulf of Mexico	1.7
sardine	Maine	1.4
tuna	California	6.2

Feedstuff	Source (if known)	Median value (mg. Se/kg. dry material)
(Fish meal)		
tuna	Samoa	5.1
whitefish	New England	1.3
Fish meal with solubles	U.S.	2.0
Fish solubles[d]	U.S.	2.0
Flour, wheat, white[c]	U.S.	0.192
Flour, whole wheat[c]	U.S.	0.634
Grass hay	Germany	0.1
Grass hay (trefoil)	New York State	0.06
Hominy feed	Midwestern U.S.	0.1
Limestone	Germany	0.17
Linseed meal	U.S.	1.0
Mangels	Germany	0.03
Mangel leaves	Germany	0.08
Meat, beef steak	U.S.; Germany	0.22
beef liver	Germany	0.3
beef heart	Germany	0.07
lamb[e]	U.S.	1.56
lamb liver[e]	U.S.	2.06
pork	Germany	0.31
pork liver	Germany	0.58
chicken[e]	U.S.	0.4
chicken liver[e]	U.S.	1.5
turkey[e]	U.S.	0.35
turkey liver[e]	U.S.	1.8
Meat and bone scraps, 50% protein	U.S.	0.29
Milk, evaporated, canned[c]	U.S.	0.012

Feedstuff	Source (if known)	Median value (mg. Se/kg. dry material)
Milk, whole dried	Germany	0.14
whole, fresh, homogenized[c]	U.S.	0.012
skimmed dried	Germany	0.14
skimmed dried	Midwestern U.S.	0.15
skimmed dried sample #1[c]	U.S.	0.096
skimmed dried sample #2[c]	U.S.	0.240
Oats	Eastern U.S.	0.05
Oats	Midwestern U.S.	0.22
Oats, steamed rolled	Midwestern U.S.	0.25
Oats, rolled	Germany	0.08
Oatmeal quick cooking[c]	U.S.	0.110
Oyster shell	Gulf of Mexico	0.01
Palm kernel meal	—	0.12
Peanuts	—	0.38
Peanut meal	—	0.28
Peas	Germany	0.27
Phosphate, dicalcium	U.S.	0.2
Phosphate dicalcium	Germany	0.21
Phosphate, rock	U.S.	1.2
Phosphate, rock	Germany	1.67
Phosphate[f]	Curaçao Island	1.2
Potatoes	Germany	0.09
Potatoes, white[c]	U.S.	0.005
Potatoes, sweet[c]	U.S.	0.007
Potato meal	Eastern U.S.	0.06
Poultry by-product meal	Eastern U.S.	1.2

Feedstuff	Source (if known)	Median value (mg. Se/kg. dry material)
Rice	—	0.13
Rye	Germany	0.20
Shrimp meal	—	1.8
Squid	—	0.46
Soybean meal	Midwestern U.S.	0.1
Soybean meal	Nebraska	0.54
Soybean meal	Eastern U.S.	0.07
Starfish meal	New Haven, Connecticut	1.3
Sugar, white[c]	—	0.003
Water, tap (mg./l.)	Ithaca, New York	0
Water, tap (mg./l.)	Stuttgart, Germany	0.0016
Water, North Sea (mg./l.)	Germany	0.0038
Wheat	Nebraska; South Dakota	0.8
Wheat	Eastern U.S.	0.05
bran	North Central U.S.	0.63
flour	North Central U.S.	0.88
middlings	North Central U.S.	0.50
middlings	Eastern U.S.	0.28
shorts	North Central U.S.	0.57
Whey, dried	Midwestern U.S.	0.08
Yeast, dried brewer's	Midwestern U.S.	1.1
Yeast, dried brewer's	Germany	0.11
Yeast, Torula	Midwestern U.S.	0.04

[a] Thompson and Scott (1968).
[b] Oelschläger and Menke (1969).
[c] Morris and Levander (1970).
[d] Sard-X, Amburgo Company, Philadelphia.

ᵉ Selenium tissue levels represent those obtained with selenium dietary levels at or slightly above the nutritional requirement.
ᶠ Curaphos, from H. J. Baker Company.

Source: Scott, M. 1971. *Poultry Science* L(6):1745.

REFERENCES

1 Frost, D.V. and Lish, P.M. 1975. *Annual Review of Pharmacology* 18:259-84.

2 *Food Technology*, 28(1):14.

3 Kubota, J. et al. 1967. *Journal of Agricultural and Food Chemistry* 15:448.

4 Patrias, G. and Olson, O. October 25, 1969. *Feedstuffs* 41:33.

5 Cunha, T. May 1973 *Feedstuffs* 45(21):48-49.

6 Morris, V. and Levander, O. 1970. *Journal of Nutrition* 100:1383-88.

7 Higgs, D. et al. 1972 *Journal of Agricultural and Food Chemistry* 20(3): 678-80.

8 Scott, M. 1973. *Journal of Nutrition* 103:803-10.

9 *Selenium in Nutrition*, National Academy of Sciences, 1971.

10 Schroeder, H. and Balassa, J. 1970. *Journal of Chronic Disabilities* 23:227-43.

11 Schrauzer, G. and White, D. 1978. *Bioinorganic Chemistry* 8:303-18.

12 Thorn, J. et al. 1978. *British Journal of Nutrition* 39:391-96.

13 Hopkins and Majaj. 1967. In *Selenium in Biomedicine*, eds. O. Muth et al. Greenwich, Connecticut: AVI.

14 Frost, D. July 31, 1971 *Feedstuffs* 43(31):12-33.

15 Passwater, R. 1973. *Fluorescence News* 7(2):11-16.

SIXTEEN

Selenium supplements

S<small>ELENIUM</small> is an element, just as carbon and oxygen are elements. Most nutrients that we eat contain carbon. Proteins, carbohydrates, fats and vitamins are all complex molecules that contain carbon. Yet, we don't speak of eating or taking our "carbon."

Minerals are single elements, but we rarely eat elemental forms of minerals. We normally eat a "salt" or an organic form of the mineral. We don't eat iron ore, we eat ferrous (iron) gluconate, ferrous sulfate etc. and we form molecules such as hemoglobin with the iron. We don't eat sodium, but we do eat sodium chloride (table salt), sodium nitrate and so on.

Each form of a mineral has its own digestibility, absorption rate and toxicity. When we consider our nutritional needs for a mineral, we must also consider the form of the mineral. This is particularly true for selenium. Because if selenium is so essential to human life, any form of selenium that is assimilable is better than none. As long as toxic amounts are not ingested, any supplement that has selenium bioavailability is better than none. But there seems to be important differences in the forms available. Some are significant, others are not. Some selenium compounds are readily utilized, others are poorly utilized. At least one selenium compound is practically non-toxic, but most are toxic, in varying degrees of toxicity. Some selenium compounds are anti-mutagenic, while others have essentially no significant activity, either anti- or pro-mutagenic. Excesses of some inorganic selenium salts can oxidize vitamin E.

Organic or inorganic

This is not a "natural" or "synthetic" debate but a discussion of the relative merits of inorganic selenium salts and organic selenium compounds normally present in foods.

Selenium in natural food sources occurs in organic forms (e.g. selenoamino acids) rather than as a mineral. The use of this organic form of selenium as a supplementary source for humans has been recommended because it is assimilated by the body with greater efficiency than are inorganic forms, and the acute and chronic toxicity of large amounts of natural selenium is substantially less than that of inorganic forms.

There is no question that both inorganic and organic forms of selenium can contribute to glutathione peroxidase and thus prevent selenium deficiency disorders. Dr. Howard Ganther of the University of Wisconsin states,

"Dietary supplements of inorganic compounds effectively prevent a variety of nutritional disorders, but there is general agreement that the active biological form of selenium is an organic form."[1]

However, not all forms of organic or inorganic selenium compounds are useful as supplements. Dr. Klaus Schwarz synthesized approximately 800 organic selenium compounds, but tests at Chemie Gruenthal showed that most tended to build up in fat tissue. The International Selenium Study Group, a subsidiary of The International Association of Bioinorganic Scientists, found that an inorganic form, selenium dioxide, failed to raise blood levels of selenium even when given in dosages as high as 1000 micrograms per day.

Natural organically-bound selenium in brewer's yeast is considerably more effective than is sodium selenite (inorganic selenium) in raising blood concentrations of this trace element. Dr. T. Westermarck reported that subjects receiving daily supplementation of inorganic selenium of 0.15 micrograms per day per kilogram body weight (1875 mcg/day for a 75 kg adult) showed an increase in blood selenium from 108 nanograms per milliliter to 141 nanograms per milliliter (a difference of 33 nanograms per milliter).[2] By contrast, Dr. Schrauzer showed when subjects received only 150 micrograms selenium per day as natural organically-bound selenium, blood selenium levels rose from 140 nanograms per milliliter to 210 nanograms per milliliter (a difference of 70 nanograms per milliliter).[3] Since a tenfold lower oral dosage of organic selenium produced a twofold greater increase in selenium levels in blood, organically-bound selenium is at least twenty-fold more effective than sodium selenite in building blood selenium levels.

In ten individuals studied, blood levels of selenium

were directly correlated with dietary selenium intakes. Activities of glutathione peroxidase in blood were not correlated with blood selenium levels. As a result, glutathione peroxidase cannot be a sufficient measure of selenium adequacy. Dietary supplementation with yeast containing organically-bound selenium (150 mcg/day) raised blood selenium concentrations from 0.14 mcg/ml to 0.21 mcg/ml over a six-week period. This rise in blood selenium was not significantly affected by concurrent zinc supplementation (15 mcg zinc/day). Two subjects with dietary selenium intakes of 350 micrograms per day and more than 450 micrograms per day, respectively, were in good health and had normal clinical blood pictures. Inorganic selenium compounds are undesirable food supplements because of their greater toxicity and poor resistance to loss during food processing. Organically-bound selenium in yeast cells is easily assimilable and of low toxicity.[3]

Dr. Duane Ullrey of Michigan State University found that selenium from natural sources had a much more pronounced effect on tissue selenium concentration than did supplemental sodium selenite. "The addition of selenium from sodium selenite to diets which were low in natural selenium sometimes produced a small increase in tissue selenium, but this value was below those found when equal dietary selenium concentrations were derived from natural sources. The addition of selenite selenium to diets which were already adequate in natural selenium produced only a small or no increase in tissue selenium."[4]

Selenium is now being considered for laying hen rations in order to increase egg production. Dr. J. D. Latshaw is currently looking at the distribution of selenium within the egg. He has found that different forms of dietary selenium fed to chickens will turn up in different parts of the egg. For instance, natural selenium in feed tends to relocate at higher levels in the whites than in the yolks. However, more selenium is found in the yolks than the whites of eggs from chickens fed sodium selenite.[5]

In a selenium symposium in Metz, France, July 9, 1979, Dr. John Martin discussed the assimilation and metabolism of organoselenium compounds.

"There are differences among living organisms concerning the metabolic pathways by which they synthesize organoselenium compounds from inorganic selenium. In general, animals, particularly mammals, first convert inorganic selenium to selenotrisulfide (1, 3-dithia-2-selanes).

"From there, much of the selenium is rather quickly excreted as either trimethyl selenonium ion in the urine or dimethyl selenide in expired air. However, some selenium must be diverted for the synthesis of at least one important selenoenzyme, glutathione peroxidase. Other selenoenzymes in animals have been implicated. However, if organoselenium compounds are supplemented to animals, they quite often follow pathways similar to that of their sulfur analogs. Selenomethionine is quite easily incorporated into the primary structure of proteins and accumulates in skeletal muscle. If the organoselenium compounds do not have sulfur analogs, they are poorly metabolized. Yet some of these compounds are reported to be remarkably active in their ability to prevent liver necrosis in rats.

"Although synthesis of selenoamino acids from inorganic selenium in animals is minimal at best, plants carry out such syntheses quite readily. Those plants which accumulate large quantities of selenium synthesize selenoamino acids whose sulfur analogs are not incorporated into proteins. Cereals and forages which do not accumulate large quantities of selenium synthesize selenoamino acids whose sulfur analogs are incorporated into the primary structure of proteins."

In an article on selenium antagonists, Dr. Gerhard Schrauzer discussed the practical aspects of dietary selenium supplementation with regards to source, dosage and safety. He concluded that selenium should be ingested in organically bound form if possible. He stated, "Ideally, selenium supplements intended for human use should contain the element in biologically incorporated form; the use of artificially selenized products or of simple inorganic selenium salts as supplements is discouraged."

Selenium yeast

My laboratory research in the late 1960s and early 1970s was limited to what was available at that time. I used many compounds, but selenocysteine was the compound of choice in my early experiments and dimethylselenide in my later experiments. These compounds are very expensive and have handling problems that seem to make them impractical for use as a dietary supplement at this time.

In 1974, a type of brewer's yeast was developed in which selenium is said to be naturally bound to the protein part of the yeast cell.

Selenium yeast appears to be the supplement of choice and is probably the most popular selenium supplement. One of its manufacturers makes the following claims:

1 Selenium yeast is biologically formed.
2 Yeast—which ordinarily contains some selenium—has been in the human food supply for many thousands of years.

3   Selenium yeast has been tested for toxicity and proven safe when used as directed.

4   Selenium yeast has been tested for biological efficacy both on humans and animals and found to be ten to twenty times as assimilable as inorganic selenium.

The manufacturer describes the product:

High selenium yeast is a food grade nutritional yeast, containing a high proportion of naturally occurring selenium. This selenium is biologically formed and organically bound into the yeast cells in the normal growth process of the yeast. The selenium is assimilated by the yeast in exactly the same manner as the small amounts of selenium which have always occurred naturally in brewer's yeast.

When living yeast are raised in a medium containing assimilable oxidized selenium salts such as selenate or selenite, this selenium is in part reduced and incorporated into selenoamino acid analogues of the sulphydryl amino acids. The pathway involved is a complex, enzymatically-mediated one, possible only within a *living* yeast cell; although the pathway is not known in detail, it is probably identical to that by which sulphate is used in sulphydryl amino acid synthesis. The resulting selenoamino acids are then incorporated into proteins in place of the analogous sulphydryl amino acids, with the result that the majority of selenium in the yeast is protein-bound.

When selenium salts are simply mixed with hydrolyzed (or autolyzed) yeast and proteolytic enzymes, there is no reason to believe that natural selenoamino acids can be formed; the situation would be completely comparable to expecting cysteine and methionine to form when sulphates are mixed with dead yeast. The forms of selenium which would result in this case would be free selenates (-nites) and an indeterminate amount of simple adducts with nucleophilic centers (especially with free cysteine).

Such adducts are by no means comparable to natural selenoamino acids, and their assimilability and efficacy have not been documented.

Dr. Klaus Schwarz found that selenium as natural selenoamino acids was about three-fold as efficient as simple oxidized selenium salts in alleviating dietary liver necrosis in rats; this results from better intestinal absorption and more efficient transfer of selenium to glutathione peroxidase. He also showed that in rats selenium in the form of selenized yeast has less chronic toxicity than do selenium salts at toxic levels. The use of natural protein-bound selenoamino acids therefore has advantages over the use of simple salts as dietary supplements.

This is not to say that selenium salts lack nutritional efficacy. But a mixture of selenium salts and their simple protein adducts is not an "organic" or "natural" form of selenium in any way comparable to the protein-bound selenoamino acid form produced by intact yeast.

I then asked Dr. Schrauzer for his views on selenium supplements. Dr. Schrauzer commented:

Selenium supplements intended for human use should contain the element in the same form or forms as it is found naturally in foods. Typically, selenium replaces sulfur in sulfur aminoacids of vegetable proteins. The naturally occurring aminoacids are predominantly in the L(+)-form. Synthetic selenium aminoacids such as selenomethionine are sold in the racemic DL-form. The biopotency of the L and D forms of selenomethionine differ. Not all naturally occurring dietary sources of selenium have been chemically identified. However, selenium is predominantly organically bound in all known dietary sources.

If a selenium supplement is to be considered a food, it is necessary to assure that the element is biologically incorporated by a living, normally growing plant or organism. All supplements containing inorganic selenium salts, or supplements to which inorganic or organic selenium compounds were added by chemical or *in vitro* enzymatic reactions such as through the use of proteolytic enzymes, are *artificially selenized* and are not acceptable sources of dietary selenium.

Suppliers of supplements containing selenium are not always clear in describing the origin of their products. Some artificially selenized products can be identified by visual inspection. In most cases, chemical tests are necessary to separate artificially selenized products from true high potency selenium yeast or other genuine natural sources of selenium.

Quantitatively, the proportion of organic and inorganic selenium in a selenium yeast product can be distinguished only by a number of more complicated tests involving extraction and chemical digestion of the cells. High potency selenium yeast contains little if any water soluble selenium, nor can much selenium be extracted on treatment with warm, 10 percent hydrochloric acid. Many of the artificially selenized yeasts will release 20-80 percent of selenium into water or warm hydrochloric acid. From genuine [selenized yeast] only 0-5 percent is released. For quantitative chemical analysis it is therefore necessary to wet-ash the yeast using a mixture of nitric and perchloric acids. If this is not done, the results for total selenium are erroneously low.

During the past three years, various commercial and experimental selenium supplements have been analyzed in our laboratory. Animal tests and experiments with human subjects have also been performed. Addition of selenium yeast to a selenium-deficient torula yeast ration prevented the out-break of selenium deficiency syndromes during twenty months (i.e. the entire life span of the mice) at levels of 0.15 ppm of selenium. A toxicity study with male Fisher 344 rats was performed by Dr. Klaus Schwarz in 1975. The data of this investigator revealed that selenium yeast is well tolerated at concentrations where inorganic selenium (selenite) shows chronic toxicity; it was estimated that selenium yeast is approximately only one-third as toxic as the latter.

Two supplementation experiments were performed in our laboratory with human subjects in 1977 and 1979. In the first study, published in *Bioinorganic Chemistry* 8: 303-18 (1978), selenium yeast was administered at levels of 150 micrograms per day for six weeks. The blood selenium levels showed a significant increase, i.e. from 0.15 to 0.21 micrograms/milliliter. Upon cessation of supplementation, the blood selenium concentrations declined by approximately 0.01 micrograms per milliliter per month; similar clearance rates were observed in studies with residents in New Zealand (N. M. Griffiths, *New Zealand Med. J.* 80: 199 (1974). The blood selenium concentrations of two subjects who had been ingesting 350 and 600 micrograms of selenium per day in form of selenium yeast were 0.35 and 0.62 micrograms/milliliter after 18 months of supplementation.

In the second selenium-supplementation study, selenium was administered to normal, healthy adults at levels of 150 micrograms per day for five weeks. The first group received selenium yeast; the second group was administered selenium in form of Complex. The third group received a 'chelated selenium yeast product.' Neither of the later two supplements produced changes of the blood selenium levels of our subjects during five weeks of supplementation. On the other hand, the original findings with a highly rated selenium yeast were fully confirmed.

The study was performed in collaboration with Mr. J. McGinness, Department of Chemistry, University of California at San Diego, and was approved by the Committee on Human Subjects, School of Medicine, University of California at San Diego.

Tests have shown that biologically incorporated selenium yeast is the preferred form of selenium supplementation for humans. Artificially selenized yeast or yeast hydrolyzates are not effective for human selenium supplementation: the inorganic selenium present is excreted from the body too quickly to allow proper utilization.[7]

Sometimes the products can be distinguished simply by checking their color. Biologically incorporated selenium yeast looks like normal yeast in every way—even under the electron microscope. Often artificially selenized yeast appears pinkish in color due to the presence of red amorphous inorganic elemental selenium. On contact with air and light, the surface turns dark red. On prolonged storage, the container may build up a sharp odor which leaves a metallic taste in the mouth and irritates mucous membranes.

If the product is very soluble in water it is most likely a yeast hydrolyzate.

The methylene blue test can distinguish biologically incorporated selenium yeast from artificially selenized yeast because it measures the availability of inorganic selenium. Biologically incorporated selenium yeast contains little, if any, inorganic selenium, while the selenium in artificially selenized yeast is essentially all inorganic or other reactive forms.

If you are not sure which type you have, you can write to the manufacturer for a copy of the methylene blue test results on his product. This is a simple, inexpensive test that all manufacturers should run. The yeast is suspended in dilute mercaptopropanol, a measured amount of methylene blue dye is added, and the time is recorded until the methylene blue solution is decolorized. A rapid decolorization indicates inorganic selenium.

Dr. Julian Spallholz's group at Texas Tech University has also studied selenium supplements. The researchers commented at a 1980 Selenium Symposium:[6]

"With both mice and rats, a commercial brewer's years enriched with selenium is being studied with respect to:

1   deposition of the selenium from the yeast into animal tissues

2   bioavailability of the selenium from the yeast for glutathione peroxidase

3   organic and inorganic selenium composition

4   ability of the selenium yeast to promote growth

"Various animal tissue samples have been analyzed for their selenium content by the fluorometric method. Deposition of selenium in animal tissues is more similar to that of animals fed diets supplemented with selenomethione than selenite. Dialysis experiments have been conducted to determine the amount of selenium organically bound in the yeast. Much of this selenium is probably bound as selenomethionine.

"These studies are considered important in view of the availability of selenium enriched yeast to the general public and the new dietary recommendations for selenium of the National Academy of Science."

As Drs. Gerhard Schrauzer and James McGinness conclude, "Only supplements containing biologically incorporated selenium can be recommended for human use. Supplements containing inorganic selenium such as sodium selenite would have to contain milligram amounts in order to cause measurable increases of blood selenium concentrations after prolonged daily administration at dosage levels of 50 micrograms per kilogram of body weight. Some commercial selenium supplements contain insoluble selenium and as such are of no apparent value."[8]

Other selenium supplements

Sodium selenite has been approved for use in animal feed and veterinary products, but has not been approved for human use by the Food and Drug Administration. Sodium selenite is less effective than selenium-yeast. "Chelated" selenium is a misnomer because selenium does not undergo chelation. Synthetic selenoamino acids are mixtures of assimilable and non-assimilable forms.

SUMMARY

Selenium scientists have proven selenium supplements are safe and effective. They recommend supplements that are found in our food supply in organic form. Other forms of selenium supplements may appear, but at this writing I can only advise you that selenium experts know more about the effectiveness and safety of biologically formed selenium-yeast than they do about other selenium supplements. The prudent approach is to go with proven forms of selenium.

REFERENCES

1 Ganther, H. In *Selenium*, eds. Zingaro, R. and Cooper, W. New York: Van Nostrand Reinhold Co., p. 546.

2 Westermarck, T. 1977. *Acta Pharmacologica et Toxicologica* 41:121-28.

3 Schrauzer, G. and White, D. 1978. *Bioinorganic Chemistry* 8:303-18.

4 Ullrey, D. May 1976. *Proceedings of the Symposium on Selenium-Tellurium in the Environment*. Pittsburgh, Pennsylvania: Industrial Health Foundation, p. 178-82.

5   *Farm and Dairy*, June 14, 1979.

6   Myers, G. et al. May 1980. Second International Symposium on Selenium in Biology and Medicine, Texas Tech University, Lubbock, Texas.

7   Schrauzer, G. et. al. 1979. *Trace Substances in Environmental Health* 13:64-67.

8   Schrauzer, G. and McGinness, J. 1980. *Journal of the International Association of Bioinorganic Scientists*, pp. 64-67.

SEVENTEEN

Toxicity

P OPULAR songwriter Ray Stevens tells us "Everything is beautiful—In Its Own Way." Let me paraphrase that, "Everything is toxic—in its own limit." Paracelsus beat me to it, of course, when he said long ago, "Dosis sola facit venenum—Only the dose makes the poison."

Perhaps selenium was neglected so long in nutritional research because it had a bum rap. Actually selenium has two bum raps—that it is toxic is true, but that it is carcinogenic is false.

All minerals, vitamins A and D, and even water and salt can be toxic. We should not fear a nutrient just because it has a toxic limit. However, we should respect that limit. The real danger of selenium is its deficiency, not its toxicity.

The toxicity range is about 100 times above the lowest effective nutrient dose. Dr. Milton Scott of Cornell University believes a maximum daily intake of 1100 micrograms of selenium daily is safe for man.[1]

In preliminary clinical studies, adults who consumed 1,000 to 2,000 micrograms of selenium (in yeast) daily for periods of over a month have failed to show any sign of toxicity.[2]

In Chapter Two, I mentioned that one physician had found no signs of nerve, liver or blood abnormalities in over 100 cancer patients given selenium supplements, often in the 900 to 2000 microgram range. This was further confirmed by thirty-seven autopsies of those patients dying of cancer.

However, this is a simplistic approach. Various selenium compounds have different toxicities. Dimethyl selenide is essentially non-toxic. Sodium selenite is more toxic than most natural occurring organic selenium compounds. As little as 3.5 parts per million of sodium selenite can produce toxicity symptoms in domestic animals.[3] This is equivalent to about 1700 micrograms per day in a human diet.

Selenium yeast is said to be only one-third as toxic as sodium selenite. Therefore the toxicity threshold of selenium yeast is about 5,000 micrograms per day.[4] This compares to the Food and Nutrition Board's safe and adequate range of 50 to 200 micrograms and the 500 micrograms of many Japanese fishermen.

Selenium has long been thought of only in terms of its toxicity. Animals consuming forage bearing 5 to 10 parts per million selenium for prolonged periods can develop selenosis commonly called "alkali disease" or the "blind staggers." The same results can occur if the animals consume forage containing up to 10,000 parts per million.[5] Of course, such toxic amounts occur only in a few high-selenium areas.

Symptoms of chronic selenium toxicity in animals are hair and hoof loss, appetite loss, nausea, stunted growth stiffness, lameness, and neurological toxicity long before death ensues. Both chronic and acute selenium toxicity produce mental confusion, "blind staggers" and a tendency to back into a corner. It is assumed that human symptoms would be similar. Only one fatality from industrial poisoning has been reported[6,7] and since man does not graze on selenium-rich forage, I doubt if any "nutritional" selenium deaths have occurred.

The FDA states, "Information with regard to selenium toxicity in man is relatively sparse, and is available only from industrial overexposure. Selenium intoxication (symptomatized by depression, languor, nervousness, and gastrointestinal disturbances) has been reported as a result of industrial inhalation, but the amount of exposure that precipitates these symptoms is not known. No data are available on human oral toxicity."[8]

Most selenium compounds are readily excreted from the body, thus reducing the possibility of toxicity unless relatively large amounts are consumed. Inorganic sodium selenite has a half-life in the body of only a few days (half of it is excreted in a few days, one-half of the remainder is excreted in another few days and so on). The level of selenium in the tissues is self-limiting at low sodium selenite intake levels.

The FDA made a point of this when they sought approval of selenium-supplemented feeds for poultry and swine: "All of the evidence available to the Commissioner shows that if the animal is already receiving a diet of natural feed with an adequate selenium level, the selenium level in meat and poultry will not be increased significantly, if at all. If the animal is receiving a diet naturally deficient in selenium, the addition of the amount of selenium proposed in this petition will raise the selenium level in the meat or poultry to no higher than the level that would be found if the feed naturally contained sufficient selenium."[8]

Lateral sclerosis

Anyone searching the medical literature for adverse effects of selenium will undoubtedly uncover a report in the *Journal of the American Medical Association* attempting to link excessive selenium intake to amyotrophic lateral sclerosis. This is a progressive degeneration of the motor cells of the spinal cord and medulla, producing atrophy, paralysis, and fasciculations (muscle contractions) of the innervated muscles.

"During a ten-year period, four cases of amyotrophic lateral sclerosis have been found in a sparsely populated county (population 4000) in west-central South Dakota. The patients were unrelated male farmer-ranchers between fifty-seven and sixty-six years of age, living within a fifteen-kilometer radius of each other. The cases occurred in a region where naturally occurring selenium toxication is endemic in farm animals."[9]

Three other "clusters" of the disease have been reported, but no link to selenium was reported in those clusters. The physicians suggest that the presence of selenium in toxic amounts in the soils of this area warrants examination of selenium as a possible environmental factor.

In summary, no cause and effect evidence exists, and persons taking supplements of selenium in the recommended nutritional range should certainly have no fear of amyotrophic lateral sclerosis.

Cancer

Chapter Two discusses the many studies that demonstrate selenium is protective against cancer. However, selenium has an unfounded stigma against it, one that has been repeated over and over since the 1940s, without justification. Early toxicity studies conducted by the FDA on a selenium containing insecticide (Selocide) produced liver tumors not because of the selenium but because the rat diet was deficient in protein which allowed cirrhosis to occur.[10] A follow-up article noted that none of the tumors metastasized, which is a requirement for classifying the tumors as malignant.[11] This false contention has gained strength by repetition.

Dr. Douglas Frost of Schenectady tells of the attempt to correct the record and how it was ignored.

"Beginning in 1959, the idea that selenium can cause cancer was questioned. Ways were sought to reevaluate that idea. This led in time to a reevaluation study at Oregon State University under contract from the National Cancer Institute. These lifetime studies in rats, with graded levels of sodium selenite and sodium selenate, found no cancers attributable to selenium. These studies carried adequate negative controls and a known liver carcinogen control as well. Discounting the conclusiveness of these findings, Drs. A. A. Nelson and M. N. Volgarev referred in the discussion of Volgarev's paper to selenium as a carcinogen. Their lead was followed by Chemical Abstracts, which also ignored the conclusions tending to clear selenium of the carcinogen charge. Oddly enough, the studies done in Russia (see Volgarev and Tscherkes) all lacked controls. The authors assigned carcinogenicity to selenium even though they showed no cancers attributable to selenium."[12]

In 1973 and 1974, the FDA Commissioner concluded:

1   The available information does not support classification of selenium or its compounds as having carcinogenic acticity,

2   The use of selenium as set forth below constitutes no carcinogenic risk, and

3   The limitations set forth below, while satisfying the animals' dietary need for selenium will assure safety to animals treated with sodium selenite or sodium selenate and to consumers of edible products of such treated animals.

For the details as to how the FDA reached this proper decision, see the following extract from the Federal Register.

The FDA rules in the Federal Register selenium does *not* cause cancer:[13]

The applicability of the anticancer clause (sec. 409(c) (3) (A) of the act to the addition of selenium to animal feed has been thoroughly considered because of the questions that have been raised concerning the possible carcinogenic activity of selenium. Available data have been evaluated by the Food and Drug Administration and the National Cancer Institute. Based on these evaluations, it has been concluded that the judicious administration of selenium derivatives to domestic animals would not constitute a carcinogenic risk. In three of the six studies available on the subject, test animals were found to have developed neoplastic lesions. These lesions were concluded to be a consequence of the liver cirrhosis produced by frank selenium toxicity. Further evaluation of the results of these three studies was complicated by the unusually high levels of selenium that had been administered, faulty experimental design, and/or infectious conditions present in the animal colonies used. Results of the remaining three studies, all of which were well controlled investigations, were negative for carcinogenic activity.

Selenium at high dietary levels (above 2′p/m for experimental animals) is a proven hepatotoxic agent. Early studies at dietary levels of 5, 7, and 10 p/m showed liver damage and regeneration in rats and an increased incidence of hepatoma in treated animals as compared with controls. Hepatoma did not occur in the absence of severe hepatotoxic phenomena. In more recent studies, hepatotoxicity was observed in rats fed selenium at 2 p/m. At 16 p/m, more severe liver damage was observed but was not associated with hepatoma. No hepatotoxic effects were noted at 0.5 p/m or below.

In this respect, selenium is no different from a number of foods and drugs available in the marketplace today. Beverage alcohol, for example, is associated with a higher incidence of liver cirrhosis, which in turn is associated with a higher incidence of liver cancer. Other common agents, at high levels, may produce the same result.

The Commissioner is of the opinion that these foods and drugs are not, by reason of their capacity to induce liver damage when abused by being consumed at high levels, properly classified as carcinogenic because of their potential association with a higher rate of liver cancer. The various anticancer clauses contained in the act (secs. 409(c) (3) (A), 512 (d) (1) (H), 706 (b) (5) (B), 72 Stat. 1786 82 Stat 345, 74 Stat. 400: 21 U.S.C. 348 (c) (3) (A), 360b(d) (1) (H), 376 (b) (5) (B) were predicated on the theory that, since we do not know the mechanisms of carcinogenesis, even one molecule of a carcinogen should not be allowed into the food supply. The anticancer clauses do not apply in the case of an agent that (1) occurs naturally in practically all foods, (2) is used in a manner such that the natural level in food is not increased, (3) has a definite hepatotoxic effect/no-effect level, and (4) has a possible carcinogenic effect which is associated only with the hepatotoxic effect.

The FDA elaborated more fully in 1974:[14]

Scientific evidence on the carcinogenicity of selenium has been published by A. A. Nelson, O. G. Fitzhugh, and H. O. Calvery, 1943, *Cancer Research* 3:230-236, H. L. Klug and C. M. Hendrick, 1954, *Proceedings of the South Dakota Academy of Science*, M. N. Volgarev and L. A. Tsherkes, 1967, *Selenium in Biomedicine* Symposium, Westport, Connecticut: AVI Publishing Co., J. R. Harr, J. F. Bone, I. J. Tinsley, P. H. Weswig and R. S. Yamamoto, same source; H. A. Schroeder and M. Mitchener, 1971 *Journal of Nutrition* 101: 1531-40, and 1972 *Archives of Environmental Health* 24:66-71.

Selenium was initially thought to be carcinogenic on the basis of the studies of Nelson et al., which were designed to compare the toxicity of graded levels of naturally occurring selenium from grains produced in seleniferous areas with that caused by potassium ammonium sulfo-selenide, a formerly used systemic insecticide. Liver tumors were produced in the treated rats, but whether or not these tumors resulted from the cirrhosis caused by the nutritionally inadequate test diets cannot be determined. While the studies of Volgarev and Tscherkes appeared to confirm the results of Nelson et al., no experimental controls were used and it was subsequently found that the experimental animals were infested with a parasite that is known to produce tumors. A third study suporting the alleged carcinogenicity of selenium conducted by Schroeder and Mitchener could not be critically analyzed, since the selenium-treated rats lived longer than the control animals, and the tumor incidence may therefore have been due to the increased life span.

A later study by Schroeder and Mitchener conducted on mice failed to show any increase in tumor incidence caused by selenium administration. The studies by Klug and Hendrick and by Harr et. al. also produced totally negative results for carcinogenic activity. These three studies were adequate, well controlled investigations in which no extraneous variables, such as those found in the studies discussed above, were present.

If you still have any doubts, please go back and read Chapter Two again. It's unfortunate that confusion exists, but misinformation is often repeated in nutrition. Don't be alarmed, however, if someday some strange selenium is found to cause cancer. But it won't be a naturally occurring organic selenium compound in the food supply. After all, we don't condemn every carbon-containing compound, just because a few are carcinogenic.

SUMMARY

Selenium is essential to human life and is safe when consumed at the levels recommended by the experts quoted in this book.

Do not experiment on your own; stay within the guidelines. Qualified physicians may suggest higher amounts during certain therapies. Even then, blood levels should be closely monitored to direct the experimentation.

REFERENCES

1 Scott, M. 1976. *Proceedings of the Symposium on Selenium-Tellurium in the Environment.* Pittsburgh, Pennsylvania: Industrial Health Foundation, p. 25.

2 Jaffe, W. Work cited above, pp. 188-93.

3 *Toxicity of the Essential Minerals,* October 1975. Washington, D.C.: Division of Nutrition, Bureau of Foods, FDA, USDHEW, p. 148.

4   Schwarz, K. December 31, 1975. 60-day toxicity trial with selenium yeast.

5   Natural Research Council. 1976. *Selenium*. Washington, D.C.: National Academy of Sciences, p. 110.

6   NRC, selenium.

7   Glover, J. 1976. *Proceedings of the Symposium on Selenium-Tellurium in the Environment*. Pittsburgh, Pennsylvania: Industrial Health Foundation.

8   April 27, 1973. *Federal Register* 38(81):10459.

9   Kilness, A. and Hochberg, F. 1977. *Journal of American Medical Association* 237(26):2843-4.

10 Nelson, A. et al. 1943. *Cancer Research* 3:220-36.

11 Fitzhugh, O. et al. 1944. *Journal of Pharmacology* 80:289-99.

12 Frost, D. July 31, 1971. *Feedstuffs* 43(31):12-33.

13 April 27, 1973. *Federal Register* 38(81):10460.

14 January 8, 1974. *Federal Register* 39(5):1355.

EIGHTEEN

Man does not live by selenium alone

Selenium is only one of the many nutrients essential to life. Our health is only as strong as the weakest link in the chain of nutrients. In many—if not most—people, that weak link is selenium.

However, taking selenium supplements while eating a poor diet only serves to strengthen one link while leaving many weak links. Selenium needs all the nutrients as partners in health, and some of these nutrients are more helpful than others. In Chapter Four, the vital roles of vitamin E and sulfur-containing amino acids were discussed in terms of their beneficial effect on selenium.

Vitamin E

The ambogenous* relationship between selenium and vitamin E has been well researched by Dr. Milton Scott of Cornell University.[1,2,3]

Vitamin E spares the body's need for selenium by its antioxidant action and also helps in the absorption of selenium compounds and vice versa. Selenium increases the activity and retention of vitamin E in the body.[4] There are some actions of each nutrient that can be performed by the other, and there are unique actions of each that the other cannot achieve.

This ambogenous relationship to health and disease is illustrated by Table 18.1, pages 210–11.

Sulfur-containing amino acids

Some of the early beneficial relationships between selenium and sulfur-containing amino acids may have been due to traces of selenium bound to those amino acids. Careful purification of the amino acids caused the apparent benefits to disappear.

However, another important interrelationship may exist. It may be that sulfur-containing amino acids such as cystine and methionine may reduce the toxicity of excess selenium. This has been debated in the scientific literature. Dr. Orville Levander has determined that vitamin E or certain synthetic antioxidants are required in order for sulfur-containing amino acids to exert their protective effect.[5]

---

*"Ambogenous" is a word used to describe two closely related nutrients, both of which are required to correct deficiency symptoms.

Eggs, by the way, are rich in sulfur-containing amino acids.

Selenium destroyers

Just as there are selenium helpers, there are selenium destroyers. Arsenic greatly increases the excretion of selenium from the body.[6] Metals such as nickel, copper, lead, cadmium, mercury, silver, thallium and zinc either interfere with the absorption of selenium or combine with selenium in the body to make it unavailable. Polyunsaturated fats use up both vitamin E and selenium as they sacrifice themselves to spare the polyunsaturates.

Avoid the metal pollutants, but make sure that you get adequate amounts of the metal nutrients such as zinc, copper and nickel. Just avoid huge unbalanced excesses of the competing minerals and have a safety margin of selenium.

REFERENCES

1 Scott, M. 1966. *Annals of the New York Academy of Science* 138:82-89

2 Scott, M. 1970. *International Journal of Vitamin Research* 40(3):334-43.

3 Combs, G. and Scott, M. 1977. *Bioscience* 27(7):467-73.

4 Desai, I. and Scott, M. 1965. *Archives of Biochemistry and Biophysics* 27(7): 467-73.

5 Levander, O. and Morris, V. 1970. *Journal of Nutrition* 100(9):1111-17.

6 Levander, O. and Baumann, C. 1966. *Toxicology and Applied Pharmacology* 9:98-115.

TABLE 18.1.
Vitamin E and selenium deficiency diseases of animals.

Disease	Animal	Tissue affected	Vit. E	Se	Anti-oxidants
REPRODUCTIVE FAILURE					
Fetal death, resorption	rat	embryonic vascular	x		x
	cow, ewe	system	x	x	
Testicular degeneration	rooster, rat, rabbit, hamster, dog, pig, monkey	germinal epithelium	x	x	
NUTRITIONAL MYOPATHIES					
Nutritional muscular dystrophy (NMD)	chick,* rat, guinea pig, rabbit, dog, monkey	striated muscle	x	†	
"Mulberry heart" disease	pig	cardiac muscle	x		
NMD	mink	striated cardiac muscle	x	x	
Gizzard myopathy	turkey	gizzard muscle	x	x	
NMD	duck	striated muscle	x		
"Stiff lamb" disease	newborn lamb	striated muscle	x	x	
NMD	sheep, goat, calf	striated muscle	‡	x	
Creatinuria	rat, rabbit, guinea pig, monkey	plasma	x		

INCISOR DEPIGMENTATION	rat	incisor enamel	‡	x	x
SYSTEMIC DISORDERS					
Liver necrosis	rat, mouse, pig	liver	x	x	x
Membrane lipid peroxidation	chick, rat	hepatic mitochondria and microsomes	x	x	x
Accumulation of ceroid §	rat, mink, calf, lamb, dog; chick, turkey	adipose	x	=	=
Kidney degeneration §	mouse, rat, pig	renal tubule contorti	x	x	x
anemia	monkey, pig	bone marrow	x	x	
ENCEPHALOMALACIA §	chick	cerebellum	x		
EXUDATIVE DIATHESIS §	chick	capillary walls	x	x	x #
PANCREATIC FIBROSIS	chick	pancreas	x	x	

*Responsive to sulfur-containing amino acids.
†May partially reduce severity.
‡Syndrome not easily produced in absence of dietary polyunsaturated fatty acids.
§Accelerated by polyunsaturated fatty acids.
||Involvement proposed but not confirmed.
#Active only in presence of selenium.

Source: Scott, M. 1970. International Journal of Vitamin Research 40(3).

NINETEEN

Perspective

SELENIUM is vital to your health. Many millions of Americans will benefit from selenium supplementation since the food supply is uncertain. That much *is* certain.

This book presents strong evidence that selenium deficiency promotes cancer, heart disease, aging, arthritis and several other diseases. This is because the selenium affects every one of your body's 30 trillion or so cells. The debate is over how much selenium is optimal for your health, not whether or not you need selenium.

Science does not always follow a predictable path. There may be future findings that shed new light on, or even invalidate, some of the evidence presented here. I have been researching selenium for nearly twenty years at this writing and urge you to consider all the evidence in this book.

You can gamble that you are getting enough selenium in your diet and that the research findings presented here are not valid. If you lose, you lose your health.

Or you can optimize your selenium intake to maximize your health.

I have recommended that you seriously consider selenium supplements. Millions of Americans are not getting the recommended dietary allowance of selenium, and millions more are not getting their optimal amount. I am concerned that the amount of selenium in our foods is decreasing as pointed out in Chapter Fifteen.

Selenium is not a panacea. It doesn't promise to cure whatever ails you. You should not entertain false hopes of a cure, just because the evidence presented here has shown that selenium can prevent or alleviate a certain disease. Selenium supplements should be used as directed, not abused. At the very least you should be aware of the important roles played by selenium and then optimize your diet accordingly.

Further research is required to elucidate selenium's role in health. In the meantime, you can sit back and watch, or you can take constructive steps now.

Best Wishes and Good Health.

Appendices

A   The free radical theory
B   Author's defense of protein missynthesis theory
C   McCarty's hypothesis on macrophage function

Appendix A

THE FREE RADICAL THEORY

Fundamentally, the free-radical theory states that in the body, the random and irreversible reactions initiated by free radicals produce a multiplicity of deleterious reactions.[1] The damage occurs from primary reactions of free radicals with DNA, RNA, RNA synthetase, and other enzymes or cell membranes, as well as by secondary damage from the chain event of lipid peroxidation.

The body has many natural producers of free radicals. Production of free radicals cannot be eliminated without destroying life itself. Besides internal sources, external sources, such as radiation, also produce free radicals.

Radiation produces free radicals by interaction with cellular water to form hydroxyl radicals (HO·), hydrogen ions, and solvated electrons (electrons associated with water molecules). The hydroxyl radical is particularly potent.

Even very small amounts of the free radicals formed by the radiolytic decomposition of water are damaging. Their devastatingly lethal effect is indicated by the fact that nuclear radiation sufficient to kill one-half of the people exposed to it produces only one free radical in every 10 million molecules exposed. Only slightly less damaging are lipid peroxidation reactions, which are about one-tenth as destructive as direct radiation. These free radicals react at a high rate with the DNA nucleotides, thymine, adenine, guanine, and cytosine. The free-radical-altered nucleotide is unstable and, in turn, reacts to alter the DNA molecule and interfere with the coding of genetic information.

Damage to DNA

Some damage to the DNA molecule can be repaired by the cell, but often it is irreparable. Usually the DNA molecule, although irreparably damaged, continues to function, albeit improperly.

Dr. Henry Eyring of the University of Utah wrote in *Science News*[2] "The free radicals produced when radiation penetrates a cell are highly reactive. They enter chemical reactions with the chromosomes of the cell and break the chromosomes, destroying some of the information carried by them. The destroyed information may range over any part of the genetic message, but sometimes it is that which controls the replication of the cell. In that case, uncontrolled replication of mutated cells may begin."

There are other mechanisms of radiation and free-radical damage to DNA. Dr. Dov Elad of the Weizmann Institute of Science has found[3] that some organic compounds can be added to the DNA molecule in such a way that they alter its function. These compounds are added by way of a free-radical mechanism.

Dr. H. J. Rhase of the National Institutes of Health also found[4] that hydrogen peroxide (H_2O_2) reacts with the DNA nucleotide adenine to modify its structure and function. The hydrogen peroxide is produced in the cells by normal body oxidation processes, but radiation is required to activate it, so that it reacts with the adenine.

It is now apparent that there is a relationship between aging and DNA damage. If the DNA molecule loses information, for example, by alteration of one or more of its nucleotides, life is impaired.

Another weak link in the body's reproduction and maintenance system is RNA synthetase. If RNA synthetase is destroyed or altered, then a particular type of RNA and DNA cannot be fabricated, and an important body function is destroyed.

This is what is involved in the aging process. Chemical reactions occur that alter RNA synthetase. The alteration results in either no protein being built by a particular RNA synthetase, or a wrong protein being built. A wrong protein inflicts damage by causing an immunological reaction, by using up required nutrients, and by strangling cells with waste material.

The alteration of RNA synthetase has been explained by scientists either as occurring solely as a function of time or by a genetic program. Research now indicates that RNA synthetase can be altered by four interrelated mechanisms: free radicals, peroxy-free-radical intermediates, oxygen dimers (a dimer is a molecule formed by joining two like molecules), and direct radiation. Oxygen taken in and stored in all living cells is capable of absorbing natural radiation (cosmic, alpha, beta, gamma, X ray, etc.), and forming the excited-state singlet dimer. The excited dimer of oxygen persists sufficiently long (>0.1 sec), and may possess sufficient energy (>8 kcal) to alter RNA synthetase and neurons through primary or secondary reactions.

Dr. Brian Stevens of the University of South Florida commented that "inhalation of singlet oxygen does provide a means for electronic excitation energy to enter the respiratory system where an important process may be the formation of a peroxide or hydroperoxide with unsaturated organic molecules, e.g., tryptophan. These contain the weak -0-0- linkage, which is more susceptible to rupture (with the formation of oxyradicals) at body temperature than most other bonds."[5]

Free-radical attack on enzymes has been widely studied. As we know, the more or less random and irreversible reactions initiated in the body by free radicals produce a multiplicity of deleterious reactions. Harman[6] listed some of these, including the following: accumulative oxidative alterations in the long-lived molecules of collagen, elastin, and chromosomal materials; breakdown of mucopolysaccharides through oxidative degradation; accumulation of metabolically inert material, such as ceroid and age pigment (lipofuscin), through oxidative polymerization reactions involving lipids, particularly polyunsaturated lipids and proteins; changes in membrane characteristics of such elements as mitochondria and lysosomes, due to lipid peroxidation; and arteriolocapillary fibrosis, due to peroxidation products of serum and vessel-wall components, probably chiefly lipids, acting as vessel-wall irritants.

Dr. A. L. Tappel[7] of the University of California at Davis stated that the knowledge of protein damage by free radicals is among the best developed. A few examples discussed by Dr. Tappel include the radiation or lipid peroxidation of proteins or enzymes in water, which results in major reactions of polymerization, polypeptide chain scission, and chemical changes in individual amino acids.

In 1941, Dr. Johan Bjorksten noticed similarities between aging human skin and the tanning of leather and protein films. He reasoned that since the latter could be controlled, then possibly human aging could be also. The chemical "tanning" is actually a "cross-linkage" of cells, where one cell is chemically united to a neighboring cell by a bridge or bond between one atom of one cell and another atom of another cell.

Dr. Bjorksten speculated that the cross-linkage could be the result of an unusual biological side reaction. He suggested that the cumulative cross-linking of body proteins is the condition known as old age.

In the June 1964 issue of *Chemistry*, Dr. Bjorksten wrote,[8] "Whether or not such cross-linking is fatal depends on the importance of the large molecules involved, how many of the cross-linkages have been formed, and in what positions."

In the early 1920s, Dr. Hermann Standinger,[9] who won the Nobel Prize in 1953 and who was a pioneer in basic work on large molecules, had proved that one single cross-link for every 30,000 components of a large molecule is enough to drastically change the solubility and behavior of the molecule. In Reference 8 Dr. Bjorksten stated, "During chemical processes within the cell, large molecules are frequently aligned side by side. This is true of DNA as it constructs other giant molecules. It is also true when enzymes act on large molecules as in digestion or synthesis, and it is particularly true when chromosomes split lengthwise in cell division, and both new chromosomes are lined up side-by-side."

In any of these instances, should free radicals appear and tie the two molecules together by cross-linking, the long molecules will be incapacitated. "Then," Dr. Bjorksten said, "many of these lead to agglomerates which cannot be resolved by any body enzyme, but will increase in the cell and gradually crowd out other constituents, thereby causing a continual decline in the cell's activity and ability to cope with stresses."

Dr. Bjorksten's theory of aging was first presented in 1942 and was independently presented[10] by Dr. Verzar in 1956 in the specific context of collagens. Since 1942, research has essentially substantiated the cross-linking theory. It has answered the ten criteria for testing theories on aging proposed by NIH's Dr. Nathan Shock in 1960.[11] In 1962, evidence was presented by Dr. Bjorksten[12] and others to prove further that cross-linking met Dr. Shock's requirements point for point and then added two new criteria to judge aging theories. Theories not incorporating the free-radical or cross-linking theories, especially somatic mutation, can answer no more than a few of Dr. Shock's criteria, let alone all of them.

The free-radical theory shows cross-linking to be caused by free radicals and adds other mechanisms. Passwater's theory adds two new mechanisms for free-radical damage, a mechanism to minimize the damage, along with considerations to extend and strengthen the cornerstone free-radical theory.

Concerning cross-linking, Dr. Tappel stated[13] as a result of recent studies that pure proteins exposed to lipid peroxidation underwent cross-linking by free-radical, chain-polymerization with resulting molecular weight distributions increasing to many multiples of that of the original protein. Tappel urged readers to "Visualize the molecular havoc of different enzymes cross-linking with their molecular neighbors in such a random destructive reaction!" Tappel also asserted, "The normal precision arrangement of proteins and enzymes in subcellular membranes and organelles would be badly disrupted, and their biological activities would be lost! . . . Another type of cross-linking of proteins which is better known to be age-related is that of maturation of collagen and elastin. However, per mole of protein affected, the free-radical-induced cross-linking would be more chemically damaging than the latter."

Recent investigations[13,14] have shed light on the damage that results from free-radical attack. When the lyosome membrane breaks, powerful hydrolytic enzymes are released that catalyze the hydrolysis of cellular components. Thus, a drastic chain-like reaction occurs that multiplies a minor event into a major problem. This amplification of the original damage explains one role that radiation plays in the aging process.

Normally, cellular fragments are hydrolyzed after being taken into the lysosomes by endocytosis. The lysosomal enzymes catalyze the hydrolysis of most of the polymeric compounds of the body proteins, poly-saccharides, and nucleic acids. However, when damaged membrane fragments are taken into the lysosomes, they are partly hydrolyzed. One result is the accumulation of age pigments with the resultant congestion of cells. Lipofuscin occupies 10-30 percent of the volume of "old" cells. It is not present in new cells. This is enough to "strangle" a cell. Another result is the temporary tie-up of lysosomal enzymes.

Observations of membrane lipid changes in aged animals have shown that the lipid content of membranes in rat liver cells of twenty-four-month-old animals is 54 percent less than that of younger ones, and that the level of a lipid-dependent enzyme, glucose-6-phosphatase, is reduced by 81 percent.

Dr. Alex Comfort remarks[14] "There is a fair amount of evidence accumulating that, while some aging processes involve gross program loss from the nuclear DNA through mutation, free-radical attack, chromosome loss, and the like, there may be more important information losses at the cytoplasmic level. The chief architect of this theory is Orgel, who suggested that random errors in RNA or its transcription into protein may produce synthetases which are both erroneous and self-replicating, with a consequent accumulation of faulty molecules and an eventual 'error crisis' leading to the death of the cell."

Most cells (skin, blood cells, etc.) are regenerated very often and the alteration of the RNA-synthetase affects the aging of these cells the most. Other cells (nerve cells, ganglia) are not regenerated. If the brain cells were regenerated, then we could not have memory storage. These cells are most subject to effects such as accumulation of lipofuscin and radiation damage.

Radiation damage can cause aging then because of the whole-body destruction of nonregenerated cells by both direct radiation and singlet-oxygen dimers, by the formation of free-radicals by direct radiation, and by the formation of peroxidized lipids via singlet-oxygen dimers.

Figure A.1 is a simplified postulated schematic of some of the factors at work in the aging process. It is important to realize that the aging process is a snowballing and self-aggravating process. The feedback mechanism involved in protein missynthesis (in which proteins are not rapidly integrated into cells due to stereochemical misfit) theoretically would cause more and more missynthesized protein to be formed, which in turn would activate the feedback mechanism more strongly, producing more and more missynthesized protein, etc. (See Appendix B.)

This is certainly a simplified picture of one aspect of the aging process.

FIGURE A-1.
Schematic of the major factors in the aging process

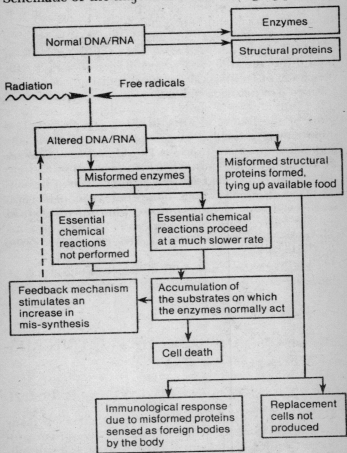

Source: Passwater. April 1971. Courtesy *American Laboratory* 3(4):36–40.

REFERENCES

1 Harman, D. 1956. *Journal of Gerontology* 11:298-300.

2 Eyring, H. 1970. *Science News* 98:270.

3 Elad, D. 1970. *Scientific American* 223(2):75.

4 Rhase, H. 1970. *Scientific American* 223(2):75.

5 Stevens, B. Private communication.

6 Harman, D. 1969. *Journal of American Geriatrics Society* 17(8):721-35.

7 Tappel, A.L. 1968. *Geriatrics* 23:97-99.

8 Bjorksten, J. 1964. *Chemistry* 37(6):6-11.

9 Standinger, H. 1937. *Zur Entwicklung der Chemie der Hockpolymeren.* Berlin: Verlag Chemie, p. 152.

10 Verzar, F. 1963. *Scientific American* 208:110-14.

11 Shock, N.W. 1960. *Aging—Some Social and Biological Aspects.* Washington, D.C.: American Academy of Arts and Sciences, pp. 250, 251.

12 Bjorksten, J. 1962. *Journal of American Geriatrics Society* 10:125-39.

13 Tappel, A.L. 1968. *Geriatrics* 23:99.

14 Comfort, A. 1970. *Geratrics* 25:3.

Appendix B

AUTHOR'S DEFENSE OF PROTEIN MIS-SYNTHESIS THEORY,

CHEMICAL AND ENGINEERING NEWS, MAY 10, 1971

SIRS: Protein missynthesis has been studied by many investigators seeking additional support for various proposed theories, including the DNA damage (somatic mutation), free-radical, cross-linking and error hypotheses.

H. J. Curtis has proposed that most of the dysfunction associated with aging may be caused by accumulative DNA damage in somatic ells, causing protein missynthesis. As malformed protein increases, an increasing number of cells function poorly or die. Curtis found that the frequency of chromosomal aberrations increased with age. Also, Curtis noted that short-lived animals accumulate aberrations faster than longer-lived animals, and that radiation causes increased aberrations in mice resulting in shortened life spans.[1,2,3] In support, H. V. Samis, Jr., et al. reported increased DNA turnover in aging rats, which they interpreted as increased DNA repairing to correct increasing damage.[4]

A. C. Allison and G. R. Paton have shown that free radicals can disrupt lysosome membranes, releasing DNases which can migrate to the nucleus and break both DNA strands with a single hit.[5]

Johan Bjorksten has proposed that DNA is a likely candidate for cross-linking which would alter its function or completely incapacitate it.[6] F. Verzar adds that the increased rigidity of cross-linked DNA may impair code transfer to messenger RNA and result in the production of nonsense protein. Verzar cites the evidence that mutagenic drugs are effective in decreasing life span and it is suggestive that they produce cross-links.[7]

L. E. Orgel has pointed out that somatic mutations, other DNA damage, and changes in repressor bonds are not the only sources of possible protein missynthesis.[8] The same factors that affect DNA can also cause mistakes in translation of DNA into RNA or translation of the nucleotide sequence of RNA into the amino acid sequence of protein. Defects in RNA polymerase, tRNA, transfer factors, amino-acyl synthetases or ribosomal subunit proteins could cause copious quantities of missynthesized proteins.

Experimentally, B. J. Harrison and R. Holliday treated Drosophila larvae with amino acid analogs and found decreased life spans. Holliday accelerated aging in fungi by administering amino acid analogs, and reportedly has extended clonal survival by fungi by 40 percent by administering antioxidants.[9,10,11]

Evidence by J. Maynard-Smith, H. V. Samis, Jr., J. A. Falzone Jr., et al. and V. J. Wulff et al. indicates that protein synthesis and RNA turnover is great in older Drosophila and mice, supporting the validity of the proposed compensation mechanism for missynthesized protein.[12,13,14,15]

Experiments have been successful in preventing free-radical damage to body components capable of mis-synthesizing proteins. Researchers feeding antioxidants to mice have greatly increased their mean life span. D. Harman reports that the percentage of mice reaching twenty months of age compared to controls has been increased sevenfold with ethoxyquin, sixfold with tert-butyl hydroxytoluene, and 30 percent with vitamin E.[16,17] A. Comfort et al. have also reported that ethoxyquin increases longevity of C3H mice.[18] N. P. Buu-Hoi reported that nordihydroguaiaretic acid increased the number of rats living to 769 days fourfold.[19] S. Ocriu and E. Vochitu reported significant life span increases with mice and guinea pigs given cysteine and other antioxidants.[20] R. Hochschild reported mean and maximum life span increases in fruit flies with antioxidants.[21] W. B. Weglicki et al. reported the accumulation of age pigment (lipofuscin) in young rats fed a vitamin E-deficient diet.[22] A. L. Tappel has reported the reduction or absence of age pigment in mice fed excess antioxidant.[23]

In addition to the twenty published articles showing possible mechanisms of production of missynthesized pro-

teins, the effect of missynthesized proteins, evidence of a
compensation mechanism for the postulated missynthesized
protein, and empirical evidence supporting mechanisms
for prevention of their occurrence, there are more
confirming reports which ease the charge that "not a
single published report of experimental evidence which
demonstrated that aging is the result of, or even accompanied by protein missynthesis." Surely the above researchers would have worked in vain if their experiments
were not considered "the slightest bit of evidence" and
they were not allowed to publish their progress toward
conclusive evidence. It is unfortunate that someone feels
that the information should be suppressed.

Many chemists in seemingly unrelated fields may have
much to offer the field of gerontology.

Research of protein missynthesis is hampered by the
difficulty in detecting the increase of an almost unlimited
kind of missynthesized protein. Increased sophistication
in instrumental methods and a greater participation by
instrumentally oriented chemists will help in our search. We
also hope to introduce instruments to speed the research.

Protein missynthesis represents only a small area of
our research in human aging.

In our attempt to find truth, we have presented at
various scientific meetings our views and plans for testing
drug formulations in humans to retard the aging process.
We have received hundreds of constructive comments
and have encouraged others to enter into gerontological
research. We applied for Investigative New Drug (IND)
exemptions with the FDA in 1970 to begin testing for
direct benefit in humans. Animal experiments performed
in the American Gerontological Research Laboratories,
Inc., and in independent laboratories under contract to
AGRL have been extremely encouraging.

REFERENCES

1 Curtis, H. and Crowley, C. 1963. *Radiation Research* 19:337-44.

2 Curtis, H. J. 1966. *Gerontologist* 6:143-49.

3 Curtis, H. J. 1967. In *Aspects of the Biology of Aging*, ed. H. W. Woolhouse. New York: Academic Press, pp. 51-63.

4 Samis, H. V. Jr., Falzone, J. A. Jr. and Wulff, V. J. 1966. *Geronotologia* 12:79-88.

5 Allison, A. C. and Paton, G. R. 1965. *Nature* (London) 207:1170-73.

6 Bjorksten, J. June 1964. *Chemistry* 37:6-11.

7 Verzar, F. 1965. *Experimental Gerontology* 1:24-31.

8 Orgel, L. E. 1963. *Proceedings of the National Academy of Science, U.S.* 49:517-21.

9 Harrison, B. J. and Holliday, R. 1967. *Nature* (London) 213:990-2.

10 Holliday, R. 1969. *Nature* (London) 221:1224-28.

11 Hochschild, R. Private communication. The author is indebted to Dr. Hochshild for several references supplied.

12 Maynard-Smith, J. 1966. In *Topics in the Biology of Aging*, ed. P. L. Krohn. New York: John Wiley & Sons, pp. 1-35.

13 Samis, H. V. Jr. 1966. *Journal of Theoretical Biology* 13:236-50.

14 Falzone, J. A. Jr., Samis, H. V. Jr. and Wulff, V. J. 1967. *Journal of Gerontology* 22:42-52.

15 Wulff, V. J., Quastler, H. and Sherman, F. G. 1962. *Proceedings of the National Academy of Science, U.S.* 48:1373-5.

16 Harman, D. 1968. *Journal of Gerontology* 23:476-82.

17 Harman, D. 1968. *Gerontologist Supplement* 8(3):13.

18 Comfort, A., Youhotsky-Gore, I. and Pathmanathan, K. 1971. *Nature* (London) 229:254-55.

19 Buu-Hoi, N.P. and Ratsimamanga, A. R. 1959. *Compt. rend. soc. biol.* 153:1180-82.

20 Ocriu, S. and Vochitu, E. 1965. *Journal of Gerontology* 20:417-19.

21 Hochschild, R. Private communication.

22 Weglicki. W. B., Reichel, W. and Nair, P. P. 1968. *Journal of Gerontology*. 23:469-75.

23 Tappel, A. L. Private communications, 1970.

Appendix C

McCARTY'S HYPOTHESIS ON MACROPHAGE FUNCTION

1. That the efficiency of macrophage participation in immunological response is relatively labile and is influenced by metabolic and pharmacological factors—this is demonstrated by immunostimulant agents which function chiefly to stimulate the reticulo-endothelial system (BCG, levamisole).

2. That the efficiency of macrophage function is energy-dependent and hinges largely upon mitochondrial function.

3. That mitochondria are particularly susceptible to structural and functional disruption by free radicals, since their one-electron transfer process can generate superoxide[1], and since they have many key oxidant-labile components, such as ubiquinone, non-heme iron proteins, and an unusually high content of polyunsaturated fats[2,3].

4. That mitochondria in macrophages are particularly subject to free-radical damage since macrophages (and granulocytes) synthesize hydrogen peroxide for bac-

teriocidal purposes[4]; intra-mitochondrial peroxide would react with superoxide to generate hydroxyl radicals.

5. That mitochondrial free-radical defense, as mediated by glutathione peroxidase (selenium dependent), the mitochondrial superoxide dismutase (Mn dependent) and vitamin E, is of the most vital importance to macrophage function, and that high levels of these factors will be found in normal macrophages.

6. That high intakes of anti-oxidant nutrients (selenium, manganese, vitamin E) or of ubiquinone, should help to optimize macrophage function and thus potentiate the immune response—immunopotentiation by each of these factors has been demonstrated experimentally[5-8].

REFERENCES

1 Noh, H. and Hegner, D. 1978. *European Journal of Biochemistry* 82:563.

2 Folkers, K. 1974. *Annual Journal of Clinical Nutrition* 27:1026.

3 Mahler, H. R. and Cordes, E. S. 1971. *Biological Chemistry*, 2nd ed. New York: Harper and Row, p. 687.

4 Roitt, I. 1977. *Essential Immunology*, 2nd ed. Oxford, England: Blackwell Scientific Publishing, p. 192.

5 Spallholz, J. E., Martin, J. L., Gerlach, M. L. and Heinzerling, R. J. 1973. *Infection and Immunity* 8:841.

6 Gontzea, I. 1974. *Nutrition and Anti-Infectious Defense.* Basel: Karger.

7 Casey, A. C. and Bliznakov, E. G. 1972. *Chemical-Biological Interaction* 5:1.

8 Bliznakov, E. G., Casey, A. C. and Premuzie, E. 1970. *Experientia* 26:953.

GLOSSARY

Adenine. A fundamental component of DNA, ATP and all nucleic acids

Ambogenous. The relationship of two closely related nutrients, both of which are required to correct deficiency symptoms

AMP. The nucleotide adenosine monophosphate

Antioxidants. Protects body cells against unwanted reactions with oxygen. Many selenium compounds are antioxidants, or anti-radicals.

ATP. The nucleotide adenosinetriphosphate, an important energy compound in metabolism

Cadmium. Pollutant mineral causing high blood pressure

Carcinogens. Substances causing cancer in various forms

Catalyst. A substance that speeds up chemical reactions but does not enter into those reactions

Coenzyme Q (Ubiquinone) An aid in certain energy-producing enzymatic reactions which are indispensible to heart function.

Cross-linking. One cell chemically unites to a neighboring cell by a bond between one atom of one cell and another atom of another cell

Cytoplasm. All living substances of a cell except the nucleus

Cytosol. Cell interior

DNA (deoxynucleic acid) and RNA (ribonucleic acid) The active stuff of a gene, bearing hereditary characteristics

Dimer. A molecule formed by joining two like molecules

Epidemiological. Survey type studies

Free radicals. Highly reactive fragments of molecules which can cause, in chain-reaction, extensive damage to cells. See Appendix A.

Hydrolysis. A splitting into simpler substances by addition of water

Interferon. An anti-viral and anti-tumor compound

Keshan disease. Potentially fatal heart disease, co-named in China: particularly affects children

Lipid peroxidation. Unneeded superoxidation in the cell

Lipofuscin. Age pigment

Lymphocytes. White blood cells formed in the lymph tissue

Lysosome. A specialized "sac" in the cell containing a mixture of hydrolytic enzymes

Macrophage. Large scavenger white blood cells

Mitochondria. Energy factories of the cell

Mutagenic. Producing gene mutations

Nucleotides. Structural units of DNA

ppm (parts per million). One molecule of a substance in with 999,999 molecules of other substances

Prostaglandin. Family of chemical messengers, much like hormones, that control many bodily functions, such as blood pressure

Selenocystine and seleno cysteine are similar selenium-containing amino acids. They are analogs of the similar sulphur-containing amino acids, cysteine and cystine.

Synergism. Action of two or more substances or organisms to achieve an effect, greater than the sums of each of the component effects

Synthetases. Enzymes that catalyze the union of two molecules

234   Selenium as Food & Medicine

Acid rain, selenium availability and, 3–4, 8–9, 10, 162–164
Adenine, 217, 232
Aging
  cell membrane and, 69, 222
  cellular, described, 66–67
  cross-linkages and, 69, 70, 219–221, 226
  DNA damage and, 68–69, 77, 217, 225
  factors in, 72–73, 221, 224
  free radical damage and, 67, 69–73, 76–77, 216–224, 225–227
  genetic vs. environmental factors and, 67–68
  lipid peroxidation and, 69, 219, 221
  lipofuscin and, 69, 78–81, 227
  lysosomal damage and, 69, 72, 221–222, 226
  mitochondria and, 71, 123–124
  protein missynthesis and, 76–77, 225–229; antioxidants and, 227; feed-back mechanism, 223; selenium and, 76
  radiation protection and, 97–98
  RNA synthetase and, 217–218, 222, 226
  superoxide dismutase and, 70–71, 231
  selenium and, 13, 63–81, 152
  vitamin E and, 152
Agricultural losses
  selenium deficiencies and, 49–51, 150–151
  soil selenium and, 139
Aiken, J., 58
Aleksandrowicz, J., 98
Allaway, W. H., 114
Allison, A. C., 226
Ambogenous, defined, 232
Amino acids
  deficiency, experimental liver necrosis and, 140
  sulfur-containing, selenium and, 147, 208
AMP, 93; defined, 230
Amyotrophic lateral sclerosis, selenium toxicity and, 200

Angina pectoris, selenium and, 7, 59, 60
Animals
  selenium deficiency and diseases, 49–51, 53, 59, 150–151, 152, 210–211
  selenium supplements for, 165–166, 186–187, 195, 199–200
  selenium toxicity and, 198–199
Antibody
  immune mechanism and, 89, 90, 91
Anticarcinogens. See Cancer
Antimutagenic activity of selenium, 40
Antioxidant(s)
  cataract and, 118
  defined, 24, 232
  dietary fats and, 32–33
  free radicals and, 24, 27, 33, 48
  life span in animals and, 65–66, 227
  lipofuscin and, 79–80, 227
  selenium and protection, 27–30
  sulfur-containing amino acids and, 208
  vitamin E and, 64, 208
Arthritis
  selenium and, 13, 84–75, 153
  vitamin E and, 153
Atherosclerosis
  heart disease and, 47
  monoclonal proliferation and, 47, 48
  selenium and, 58–59, 60, 250–262
Atkinson, D., 119–120
ATP, defined, 232

Badiello, R., 97
Baumann, C., 24, 27
Benditt, E., 48
Berenshtein, T. E., 92–93
Birth rate, selenium and, 125, 127
Bittner-virus induced tumors, 133–134
Bjorksten, J., 7, 54–56, 65, 219, 220, 226
Black, M. M., 133
Blood pressure, selenium and, 57–60
Blood selenium
  assay, 133
  cancer and, 18, 21, 22, 41, 42, 43

organic vs. inorganic supplementa
tion and, 185–186
Bosco, D., 98
Boynton, H. H., 4
Buu-Hoi, N. P., 227

Cadmium
defined, 58, 230
poisoning, 13, 129–130
Calcium metabolism, vitamin E and
selenium and, 153
Cancer. See also Carcinogen(s)
age and, 64
blood levels of selenium related to,
18, 21, 22, 41, 42, 43
cell growth in, 24
chemically induced, 24
chemotherapy, adjunctive selenium
and, 42–43
crop selenium related to, 19, 20
dietary fat and, 34–38
dietary selenium and, 6–7, 14–43,
52, 133–134, 136, 137, 152
dietary selenium and fat and, 32–33
dietary selenium supplements in
protection against, 30–33, 43
energy sources and, 27
free radicals and, 69
incidence distribution by states, 11
Schrauzer on selenium and, 133–
134
selenium antagonists and, 134
soil selenium and, 19, 20, 21, 24, 93
survival time, selenium and, 41–43
Cannon, H., 52
Carcinogen(s)
active vs. pre-carcinogen, 30
defined, 27, 232
liver and, 40–41
selenium as, 201–205
selenium in protection against, 27–
30, 36–40, 43
Carter, D., 49
Castillo, R., 109
Catalyst, defined, 232
Cataract
causes, 118–119, 121
metabolic factors and, 120–121
selenium and, 13, 119, 121, 152

vitamin E and, 152
Cell membrane
free radical damage to, 69, 222
organelles and selenium, 69, 222
Clayton, L., 24, 27
Coenzyme Q (ubiquinone)
defined, 230
energy metabolism and, 124
heart function and, 57
immune system and, 92, 124
muscular dystrophy and, 114–117
selenium and, 92, 93, 117, 152
vitamin E and, 152
Colombetti, G., 97
Comfort A., 222, 227
Coronary thrombosis, 222, 227
Cowgill, U. M., 125
Crops. See Plant selenium
Cross-linkages
free radicals and aging and, 69, 70,
219–221, 226, 232
Cunha, T., 166
Curtis, H. J., 226
Cystic fibrosis, 100–112
dietary management, 108–11, 112
as genetic defect, 101, 102
maternal selenium deficiency and,
100, 103–106, 107
organ changes and selenium, 101,
102, 103
selenium therapy, 108, 111–112
vitamin E therapy, 111
Cytoplasm, defined, 232
Cytosol, defined, 232

Daft, F. S., 142
Dayton, S., 34
Detoxification of heavy metals
selenium in, 13, 128–131, 151, 152
vitamin E in, 152
Diet(s). See also Dietary selenium; Dietary selenium supplementation
experimental liver necrosis and,
138, 140–150
fat: cancer and, 32–33, 34–36, 36–
38; saturated vs. polyunsaturated,
33, 34; selenium and, 30, 32–33
Dietary selenium. See also Dietary selenium supplementation

adequate, 158, 161
animal deficiencies, 49–51, 53, 59, 150–151, 152, 210–211
availability in foods, 167
birth rate and, 125, 127
cancer and, 6–7, 14–43, 52, 133–134, 136, 137, 152; risk and, 18, 43; survival time, 40–43
estimates of typical, 167–168
heart disease and, 53–56, 59
heavy metal poisoning and, 128, 130–131
lipofuscin and, 80
maternal, cystic fibrosis etiology, 100, 103–106, 107
sources and levels in foods, 135, 137, 170, 171–182
Dietary selenium supplementation. *See also* Recommended daily allowances (RDA) of selenium
in animals, 165–166, 186–187, 195, 199–200
in cancer protection, 30–33, 43
in cystic fibrosis, 108–111, 112
guidelines, 9, 12, 205, 213
needs for, 10–12
organic vs. inorganic, 184–188
studies of, 191–192, 193–194
toxicity and, 108–109, 197–198
yeast in, 185–186, 188–190, 192–193
Dimer, 218, 232
Dinning, J., 114–115
DNA
aging and, 68–69, 77, 217, 226
defined, 232
free radicals and, 68–69

Elad, D., 215
Environmental stress, selenium and, 135
Enzymes reactions, free radicals and, 68, 218–219, 221
Epidemiology, defined, 232
Epstein, P., 111
Eriksen, N., 133
Eyring, H., 217

Factor 3

experimental liver necrosis and, 138, 142–143, 144–147
structure/activity relations, 149
Factor G., 138
Factor H., 140
Farrell, P., 111
Fat(s)
dietary: cancer and, 32–38; saturated vs. polyunsaturated, 33, 34; selenium and, 30, 32–33
as selenium destroyers, 209
Fertility, selenium and, 125–126, 127
Fertilizers,
plant selenium and, 3–4, 8–9, 10
soil selenium and, 162, 163, 166
Fielden, M., 97
Flohé, L., 71, 148
Folkers, K., 115
Foltz, C., 138
Fredga, A., 148–149
Free radicals
aging and, 67, 69, 70–71, 72, 73, 76–77, 216–224, 225–226, 227
antioxidants and, 24, 27, 33, 48
cancer and, 69
cell membrane damage and, 69, 222
defined, 25, 68, 232
cross-linkages and, 69, 70, 219–221, 226
enzyme metabolism and, 68, 218–219, 221
glutathione peroxidase and, 58–59
heart disease and, 69
hydrogen peroxide and, 69, 217, 218
lipid peroxidation and, 69, 219, 221
lysosomal damage and, 69, 72, 221–222, 226
mitochondrial damage and, 71, 123–124, 228
production of, 25–27, 33; natural, 214
radiation damage and, 68–69, 217
RNA synthetase and, 217–218, 222, 227
superoxide dismutase of, 70–71, 231
Frost, D., 5, 16, 18, 52, 92, 163, 169, 201; overview of research of, 151–152

Gamma rays, selenium protection
    against, 98
Ganther, H., 184–185
Garner, M., 119
Glutathione peroxidase
    free radical actions and, 58–59
    functions, 70–71
    metabolism, 74
    mitochondrial damage and, 123, 231
    selenium and, 119, 137, 148, 151,
        184, 186
    vitamin E and, 151
Goddard, R. F., 110–111
Godwin, K. O., 48–49
Good, R., 89
Greeder, G., 40
Griffin, C., 28–29, 29–30, 38–39
Gullino, P., 16
Gunn, S., 125
Gyorgy, P. 141

Harman, D., 64, 69, 217
Harr, J. R., 27–28
Harrison, B. J., 227
Hartley, W., 124–125
Heart disease
    angina pectoris, 7, 59, 60
    animal, 49–51, 53, 59
    atherosclerosis and, 47
    coenzyme Q and, 57
    free radicals and, 69
    Keshan, 7–8, 51, 56–57, 60, 152
    myocardial infarction and, 46–47
    selenium and, 7–8, 12, 46–59, 60,
        137, 152
    vitamin E and, 152
    Zenker's disease, 48, 49, 50
Heavy metals
    detoxification: selenium and, 13,
        128–131, 151, 152; vitamin E and,
        152
    as selenium antagonists, 161, 209
Hegner, R., 71
Hidiroglou, M., 116
Himsworth, A., 141, 142
Hochschild, R., 227
Hodgson, J. F., 114
Hoekstra, W., 119, 130, 148, 151
Holliday, R., 227

Hydrogen peroxide, free radical reac-
    tions and, 69, 217, 218
Hydrolysis, defined, 232

Immune system, 88–95
    coenzyme Q and, 92, 124
    major components and mechanisms,
        89–92; antibodies, 90, 91;
        macrophages, 92, 230–231
    selenium and, 12, 41, 88–89, 92,
        93–94, 231
Impotence, 124
Infection resistance, selenium and, 89
Interferon, 89, 90, 92, 232

Jacobs, M., 28–29
Jansson, B., 36

Keshan disease in children, selenium
    and, 7–8, 51, 56–57, 60, 152
Kubota, J., 165
Kuhn, R., 139

Lane, H., 38–39
Latshaw, J. D., 186
Lead poisoning, 13, 129
Legionnaires' disease, selenium defi-
    ciency and, 94
Levander, O., 5, 129, 167, 208
Life span
    antioxidants and, 65–66, 227
    selenium and, 65, 81
Lillie, R. D., 142
Lipid peroxidation
    defined, 232
    free radicals and, 69, 219, 221
Lipofuscin
    aging and, 69, 78–81, 223
    defined, 232
    selenium and antioxidants and, 78–
        81, 227
Liver
    Schwarz's work on experimental
        necrosis, 138, 140–150
    selenium and function of, 40–41
Lymphocyte(s)
    defined, 89, 232
    immune system and, 89–90, 91, 92
Lysosome

defined, 233
free radicals and damage to, 69, 72, 221–222, 226
selenium and, 136
vitamin E and, 136

McCarty, M., 92, 230–231
McConnell, K. P., 41
McDade, J. E., 94
McGinness, J., 192, 194
Mckay, R., 111
Macrophage(s), 91
  defined, 89, 233
  immune response and, 92, 230–231
  interferon production and, 90, 92
  mitochondria and, 92, 93–94
  selenium and production of, 92, 230–231
Martin, J., 93, 187
Martin, W., 116
Mason, K. E., 144
Maynard-Smith, J., 227
Mercury poisoning, 13, 128, 130
Mertz, W., 138, 139, 156
Metastatic breast cancer, selenium and, 15
Metals. See Heavy metals
Milner, J., 39–40
Mitochondria
  defined, 7, 123, 231
  free radical damage, 71, 123–124, 228; glutathione peroxidase and, 71, 123, 231
  macrophages and function of, 92, 93–94
  results of impaired function, 123–124
  selenium and function of, 7, 127
Monoclonal proliferation, atherosclerosis and, 47, 48
Muscular dystrophy
  coenzyme Q and, 114–115, 116, 117
  selenium and, 47–48, 113–117
Mutagenic
  antimutagenic activity of selenium, 40
  defined, 233
Myocardial infarction, 46–47

Myopathy, selenium and vitamin E deficiency and, 210

NADH, cystic fibrosis and, 101
Nelson, A. A., 201
Nohl, P., 71
Nucleotide, 217, 233
Nutrition, arthritis and, 84

Ocriu, S., 227
Olson, O., 165
Orgel, L. E., 226

Paton, G. R., 226
Pearce, M., 34
Periodontal disease, vitamin E and selenium and, 152
Perry, H., 58
Plant selenium, 8–9, 10
  acid rain and, 3–4, 162–164
  cancer and, 19, 20
  fertilizers and, 162, 165, 166
Prostaglandins
  defined, 7, 57, 233
  selenium and 7, 57–58
Protein missynthesis theory of aging, 76–77, 225–229
  antioxidants and, 227
  feedback mechanisms and, 223
  selenium and, 76
Pryor, W., 68–69

Radiation
  antioxidants and, 97–98
  "first aid" kit against, 96–97
  free radical damage and, 68–69, 216–217
  selenium compounds and, 97–98
  superoxide dismutase and, 97
Reproduction, vitamin E/selenium deficiency and animal, 210; See also Sexual function
Recommended daily allowances of selenium, 9, 12, 109, 155–159; See also Dietary selenium supplementation
  FDA and, 157–158
  optimal, 169
  problems in establishing, 155–157

range, 167
Shamberger on, 136, 158
Schrauzer on, 134–135, 158
tolerance to, 152
toxicity and, 108–109, 135, 156, 157, 197–198
Rhasé, H. J., 215
RNA synthetase, free radical reactions in aging and, 217–218, 222, 227
Roberts, N., 90, 92
Roth, G., 77
Rotruck, J., 148

Samis, H. V., Jr., 224
Schmitz, J. A., 150
Schrauzer, G., 5, 6, 15–16, 30, 42, 132, 161
on dietary selenium intakes, 167–168
on dietary selenium supplementation, 188, 190–192, 194
overview of research by, 133–135
on selenium dosages, 185
Schroeder, H. A., 129–130, 167
Schultz, R., 93–94
Schwarz, K., 5, 40–41, 49, 132, 151, 185
on selenium supplementation, 190
overview of research by, 137–151
Scott, M., 5, 40–41, 49, 64, 166–171, 197, 208
Sebrell, W. H., 142
Selenoamino acids
described, 233
as dietary supplements, 187–188
radiation protection and, 97, 98
Sexual function, 13, 123–127
malnutrition and, 124–125
selenium and, 13, 124–126, 127
Shamberger, R., 5, 7, 18, 24, 34, 52, 53, 54, 58, 132; overview of research by, 136–137
Shapiro, B., 101
Shaw, C., 16, 18, 30
Sheffy, B., 93–94
Shock, N., 220–221
Sodium selenite, cataract and, 121
Soil selenium, 8, 9, 10
acid rain and, 3–4, 8–9, 10, 162–163

agricultural losses and, 139
cancer incidence and, 19–21, 24, 93
distribution, 162, 164
factors in depletion, 162–166
fertilizers and, 162, 165, 166
Spallholz, J., 93, 193–194
Spector, A., 119
Sperm, selenium deficiency and, 125–126, 127
Sprinker, L. H., 59
Standinger, H., 220
Sterility
incidence, 124
selenium and, 125–126, 127
Stevens, B., 218
Stich, H., 40
Stress
arthritis and, 84
selenium protection against, 135
Sulfur-containing amino acids, selenium and, 147, 208
Superoxide dismutase (SOD)
arthritis and, 85
free radical actions and, 70–71, 231; mitochondria, 231
radiation protection and, 97
Supplementation. See Dietary selenium supplementation; Recommended daily allowances of selenium
Survival time in cancer, selenium and, 41–43
Synergism, defined, 233
Synthetase
defined, 233
RNA: free radical actions and, 217–218, 222, 227; functions of, 217–218

Tappel, A., 5, 65, 69, 77, 78–79, 97, 219, 221, 227
Tetanus toxoid, selenium and antibody response to, 93
Tolsem (Chromalloy), clinical trial in angina, 59
Toxicity of selenium, 136–137, 169, 197–205
amyotrophic lateral sclerosis and, 200

to animals, 198–199
carcinogenicity, 201–205
compound related to, 149–150, 198
dosage and, 135, 156
excretion related to, 199
Schwarz's research on, 138–139

Ubiquinone (coenzyme Q)
defined, 232
energy metabolism and, 124
heart function and, 57
immune system and, 92, 124
muscular dystrophy and, 114–117
selenium and, 92, 93, 117, 152
vitamin E and, 152
Ullrey, D., 186

Verzar, F., 220, 226
Vincent, J. E., 58
Vitamin A in cataract therapy, 120
Vitamin C in cataract therapy, 120
Vitamin E
aging and, 152
animal deficiency disease, 210–211
antioxidants and, 64, 208
calcium metabolism and, 153
catalytic functions of, 148
cataract and, 118, 119, 121
coenzyme Q synthesis and, 152
in cystic fibrosis management, 111
experimental liver necrosis and, 140,
142, 143–144, 146–147
heart disease and, 59, 152
heavy metal detoxification by, 129,
151, 152
immune response and, 92, 229
importance of, 52
lead poisoning and, 129
life span of animals and, 227
lipofuscin and, 79
lysosomal membrane and, 136
mitochondria and, 124, 231

in muscular dystrophy therapy, 113–
117
myopathy and deficiency in animals,
210
periodontal disease and, 152
selenium functions related to, 64–
65, 147–148, 184, 208
selenium plus: in aging, 152; in
animal myopathies, 210; in arthritis, 84, 86, 153; in cataract, 152;
glutathione peroxidase and, 151;
in heart disease, 59, 152; in heavy
metal detoxification, 151; in muscular dystrophy, 113–117; in
periodontal disease, 152
sexual function and, 124
sulfur-containing amino acid function and, 208
Vochitu, E., 227
Volgarev, M. N., 201

Wallach, J., 100, 101, 103–105, 108,
109
Wattenberg, L., 28
Warburg, O., 27
Weglicki, W. B., 225
Wei, L., 40
Weichselbaum, T. E., 142
Westermarck, T., 185
Whanger, P. D., 30, 126, 130
Whiting, R., 40
Williams, R., 120, 158
Willis, C., 53
Wulff, V. J., 225

Yeast
for selenium supplementation, 185–
186, 188–190, 192–193
toxicity, 198

Zenker's heart disease, 48, 49, 50